AF877310

BIOLOGY OF STRESS IN FARM ANIMALS:
AN INTEGRATIVE APPROACH

CURRENT TOPICS IN VETERINARY MEDICINE AND ANIMAL SCIENCE

Control of Reproduction in the Cow, edited by J.M. Sreenan

Patterns of Growth and Development in Cattle, edited by H. de Boer and J. Martin

Respiratory Diseases in Cattle, edited by W.B. Martin

Calving Problems and Early Viability of the Calf, edited by B. Hoffmann, I.L. Mason and J. Schmidt

The Future of Beef Production in the European Community, edited by J.C. Bowman and P. Susmel

Diseases of Cattle in the Tropics: Economic and Zoonotic Relevance, edited by M. Ristic and I. McIntyre

Control of Reproductive Functions in Domestic Animals, edited by W. Jöchle and D.R. Lamond

The Laying Hen and its Environment, edited by R. Moss

Epidemiology and Control of Nematodiasis in Cattle, edited by P. Nansen, R.J. Jørgensen and E.J.L. Soulsby

The Problem of Dark-Cutting in Beef, edited by D.E. Hood and P.V. Tarrant

The Welfare of Pigs, edited by W. Sybesma

The Mucosal Immune System, edited by F.J. Bourne

Laboratory Diagnosis in Neonatal Calf and Pig Diarrhoea, edited by P.W. de Leeuw and P.A.M. Guinée

Advances in the Control of Theileriosis, edited by A.D. Irvin, M.P. Cunningham and A.S. Young

Fourth International Symposium on Bovine Leukosis, edited by O.C. Straub

Muscle Hypertrophy of Genetic Origin and its Use to Improve Beef Production, edited by J.W.B. King and F. Ménissier

Aujeszky's Disease, edited by G. Wittman and S.A. Hall

Transport of Animals Intended for Breeding, Production and Slaughter, edited by R. Moss

Welfare and Husbandry of Calves, edited by J.P. Signoret

Factors Influencing Fertility in the Postpartum Cow, edited by H. Karg and E. Schallenberger

Beef Production from Different Dairy Breeds and Dairy Beef Crosses, edited by G.J. More O'Ferrall

The Elisa: Enzyme-Linked Immunosorbent Assay in Veterinary Research and Diagnosis, edited by R.C. Wardley and J.R. Crowther

Indicators Relevant to Farm Animal Welfare, edited by D. Smidt

Farm Animal Housing and Welfare, edited by S.H. Baxter, M.R. Baxter and J.A.D. MacCormack

Stunning of Animals for Slaughter, edited by G. Eikelenboom

Manipulation of Growth in Farm Animals, edited by J.F. Roche and D. O'Callaghan

Latent Herpes Virus Infections in Veterinary Medicine, edited by G. Wittmann, R.M. Gaskell and H.-J. Rziha

Grassland Beef Production, edited by W. Holmes

Recent Advances in Virus Diagnosis, edited by M.S. McNulty and J.B. McFerran

The Male in Farm Animal Reproduction, edited by M. Courot

Endocrine Causes of Seasonal and Lactational Anestrus in Farm Animals, edited by F. Ellendorff and F. Elsaesser

Brucella Melitensis, edited by J.M. Verger and M. Plommet

Diagnosis of Mycotoxicoses, edited by J.L. Richard and J.R. Thurston

Embryonic Mortality in Farm Animals, edited by J.M. Sreenan and M.G. Diskin

Social Space for Domestic Animals, edited by R. Zayan

The Present State of Leptospirosis Diagnosis and Control, edited by W.A. Ellis and T.W.A. Little

Acute virus infections of poultry, edited by J.B. McFerran and M.S. McNulty

Evaluation and Control of Meat Quality in Pigs, edited by P.V. Tarrant, G. Eikelenboom and G. Monin

Follicular Growth and Ovulation Rate in Farm Animals, edited by J.F. Roche and D. O'Callaghan

Cattle Housing Systems, Lameness and Behaviour, edited by H.K. Wierenga and D.J. Peterse

Physiological and Pharmacological Aspects of the Reticulo-Rumen, edited by L.A.A. Ooms, A.D. Degryse and A.S.J.P.A.M. van Miert

Biology of Stress in Farm Animals: An Integrative Approach, edited by P.R. Wiepkema and P.W.M. van Adrichem

BIOLOGY OF STRESS IN FARM ANIMALS: AN INTEGRATIVE APPROACH

A seminar in the CEC programme of coordination research on animal welfare, held on April 17–18, 1986, at the Pietersberg Conference Centre, Oosterbeek, The Netherlands

Edited by

P.R. Wiepkema
Department of Animal Husbandry-Ethology, Agricultural University Wageningen, The Netherlands

P.W.M. van Adrichem
Department of Animal Physiology, Agricultural University Wageningen, The Netherlands

1987 **MARTINUS NIJHOFF PUBLISHERS**
a member of the KLUWER ACADEMIC PUBLISHERS GROUP
DORDRECHT / BOSTON / LANCASTER
for
THE COMMISSION OF THE EUROPEAN COMMUNITIES

Distributors

for the United States and Canada: Kluwer Academic Publishers, P.O. Box 358, Accord Station, Hingham, MA 02018-0358, USA
for the UK and Ireland: Kluwer Academic Publishers, MTP Press Limited, Falcon House, Queen Square, Lancaster LA1 1RN, UK
for all other countries: Kluwer Academic Publishers Group, Distribution Center, P.O. Box 322, 3300 AH Dordrecht, The Netherlands

Library of Congress Cataloging in Publication Data

Biology of stress in farm animals.

(Current topics in veterinary medicine and animal science)
Includes bibliographies and index.
1. Veterinary physiology--Congresses. 2. Stress (Physiology)--Congresses. 3. Livestock--Behavior--Congresses. 4. Stress (Psychology)--Congresses.
I. Wiepkema, P. R. (Pieter Reinier), 1928- .
II. Adrichem, P. W. M. van. III. Commission of the European Communities. IV. Series.
SF768.B54 1987 636.089'689 87-11326
ISBN 0-89838-895-3 (U.S.)

ISBN 0-89838-895-3
EUR 10604 EN

Copyright

© 1987 by ECSC, EEC, EAEC, Brussels-Luxembourg.

All rights reserved. No part of this publication may be reproduced, stored in a retrieval system, or transmitted in any form or by any means, mechanical, photocopying, recording, or otherwise, without the prior written permission of the publishers,
Martinus Nijhoff Publishers, P.O. Box 163, 3300 AD Dordrecht, The Netherlands.

PREFACE

This book contains the contributions to a workshop on stress in farm animals held on April 17-18, 1986 at the Pietersberg Conference Centre Oosterbeek. The workshop was financed by the Commission of the European Communities from its budget for the coordination of Community Agricultural Research and the Agricultural University of Wageningen (The Netherlands). Its aim was to bring together experts from different disciplines all having in common that in one way or another they were involved in stress research. Such a multidisciplinary encounter should not only provide an interesting description of present day knowledge on stress, but also promote a more integrated view on stress phenomena as they occur in higher vertebrates. In the course of this workshop the following fields of research were related to stress: endocrinology, immunology, pathology, neurobiology, ethology and theoretical biology. Each of these relationships was introduced by one speaker presenting a concise state of the art. The same relationship was elaborated by a second speaker implementing the available knowledge as far as possible to the farm animal situation. Therefore in this book each discipline is represented by a duo and introduced by some integrating remarks.

This design facilitated the exchange of relevant and often unknown data among experts stemming from quite different fields of research and interest. A significant fact was that many of the attendants met each other for the first time and often had not even heard from each other before. This book might help in bridging gaps not only by offering the text of the different contributions, but also by containing so many references about stress research in different disciplines.

Of course a central question in most discussions was the meaning of the concept of stress itself. Although it was accepted that the concept especially refers to those forms of interaction of the individual, in which the latter is more or less overtaxed, an adequate answer is still in a developmental stage. This is partly caused by the existence of an impressive professional jargon. For the main part, however, it has to do with our lack of proficiency to integrate really different view-points and data into one and the same model. That appears to be the price specialists have to pay for the fascinating progress they often realize in their own terrain.

As could be expected in our discussions, the concept had not always the same contents. As a rule, however, it referred to the individual response state evoked by stressors. Thanks to the pioneering work of Mason (1971) and others we now know that such stressors are not only physical ones - as emphasized by the founder of stress research Selye (1936) -, but also psychological ones. In fact the idea that the psychological aspects of stressors are crucial is rapidly gaining ground. These neurobiological and cognitive aspects of stress phenomena dominated the discussions of the present workshop. If these aspects are really crucial, they have an enormous impact on our thinking about animal health and welfare questions in present day husbandry systems. For stress, and especially chronic stress, may significantly reduce profitable and therefore relevant traits of farm animals, for instance, by reducing the efficiency of metabolism, by facilitating the performance of disturbed behaviour or by decreasing reproductive capabilities and the resistance against pathogens. Therefore one should minimize stress as much as possible. To realize this we cannot overlook cognitive and emotional processes in the organisms involved.

We hope that the present reflection of the workshop on stress in farm animals may stimulate and integrate research on this most relevant and fascinating field.

REFERENCES

Mason, J.W., 1971. A re-evaluation of the concept of "non-specificity" in stress theory. J. Psychiat. Res. 8, 323-333.

Selye, H., 1936. A syndrome produced by diverse nocuous agents. Nature, 138, 32.

P.R. Wiepkema
P.W.M. van Adrichem
January 1987

CONTENTS

INTRODUCTION

We have passed the stage in which the idea prevailed that the main endocrinological response to a stressor was reflected in an enhanced activity of the pituitary-adrenocortical system. Closely and crucially connected with this system is the hypothalamus, forming the so called hypothalamic-pituitary-adrenal axis. The involvement of the hypothalamus also implies an important contribution of the autonomic nervous system comprising the sympathetic and the parasympathetic divisions. The development of reliable methods to measure minute quantities of hormones and of neurotransmitters produced such a complex endocrinological picture, that at this moment it is difficult to state which is the central or main endocrinological response during stress.

As described by Oliverio, there is a most intricate pattern of endocrinological activities evoked by ordinary stressors. The picture is still more complicated, as indicated by Ladewig, in that a number of hormonal reactions change significantly over time and, moreover, may differ from one individual to another. Apart from this, endocrinological responses also depend on circadian rhythms existing in all the organisms involved.

All this leads to the bewildering conclusion that neuroendocrinological responses to stressors are so complex both in patterns of participating hormones and in changes over time, that final insight seems far out of reach. Although it is useful not to overestimate scientific perspectives, we should not adhere to the reverse. If we correct for daytime and individual factors, a most important cue is without doubt that most if not all endocrinological events typical during stress do not occur independently. They interact in a well-structured way. A main principle appears to be the occurrence of feedbacks, be they negative or even positive. Analysis of these feedback patterns may significantly improve our understanding of how the perplexing puzzle of hormonal changes during stress can be fitted. As indicated by Oliverio, a crucial role has been reserved for the brain in controlling the endocrine system (see also the contribution of Warburton); a specific role is attributed to neuropeptides, among which the corticotrophin releasing factor (CRF) attracts special attention.

We not only have to know more about the highly complicated interactions in the neuro-endocrine system (in which the role of the parasympathetic system sometimes seems to be underexposed, cf. the chapter of Bohus et al.), but we also urgently need research aiming at the biological significance of these endocrine interactions for an intact individual coping with changes in its normal environment. No doubt these endocrinological processes underlie relevant homeostatic aspects of the organism's life. However, these aspects not only imply classical ones like the regulation of body temperature or of glucose availability and others, but also those that govern homeostatic aspects of social organization and of controllability of environmental changes. This latter point is well illustrated in a scheme derived from Henry, indicating the relationship between hormonal patterns during stress and controllability of the environment. Next to these homeostatic effects, the change in hormone profiles will also influence the control in the partitioning of nutrients (homeorhetic control).

The emerging picture indicates that also in farm animals during stress very complicated hormonal changes take place. Since in this type of animals so-called chronic stress is common, it is useful to remember Ladewig's remarks on chronic stress. He points out that contrary to what the term suggests chronic stress need not imply a continuous state of being overtaxed and upset. Although the stressors may be present continuously, the animal's response state needs not be a continuous one. Apart from habituation, specific functional states (for instance, rest and sleep) may significantly interrupt chronic stress.

ENDOCRINE ASPECTS OF STRESS: CENTRAL AND PERIPHERAL MECHANISMS

Alberto Oliverio
Istituto di Psicobiologia a Psicofarmacologia, C.N.R.,
1 via Reno 00198 Rome Italy
and
Dipartimento di Genetica e Biologia Molecolare, University of Rome

ABSTRACT
The relationship between the hypothalamic - pituitary - adrenocortical (HPA) and the sympathetic - adrenomedullary (SA) systems is discussed by considering their role at the central and peripheral level. HPA and SA responses are not independent from each other and implicate a number of hormonal, monoaminergic and peptidergic reactions at the brain level. It is indicated that a number of interacting systems and feedback mechanisms take part in stress responses and that the corticotrophin releasing factor (CRF) has a major role at all levels of the neuroendocrine effector system.

A number of neurochemical responses occur during stress. Within these responses two systems have been considered in detail, the hypothalamic - pituitary - adrenocortical (HPA) system, and the sympathetic - adrenomedullary (SA) system. As indicated by Dunn and Kramarcy (1984) most experiments have been focused on the one or the other of these systems while it is more and more evident that both are biologically significant and that a number of critical interactions occur between the HPA and SA systems. Similarly, a number of recent data indicate that as it is meaningless to consider the HPA and SA systems as two separate aspects of stress reactions, it is also evident that a number of peptides modulate stress responses at the central and peripheral level by acting on neural, endocrine and immune systems, thus defining a group of cells and reaction patterns whose function may be to integrate information through a psycho-immunoendocrine network.

The first studies on the response of the organism to stressful situations were those of Cannon and de la Paz (1911) who showed that the adrenal medulla and the SA system were involved in emergency situations. Cannon noted that the venous blood of cats frightened by barking dogs contained adrenaline, a response of the organism which was prevented by adrenalectomy or by section of the splanchnic nerve innervating the adrenal medulla (Hitchings et al., 1913). Cannon suggested that the adrenal medulla was acting in concert with the sympathetic nervous system, so that both systems were activated during stress. The role of the sympathetic nervous system in response to stressful events was later emphasized by the experiments of Mason (1968) and Maickel et al (1967): these authors clearly showed that stressors activate the sympathetic system causing it to release noradrenaline (NA), and the adrenal medulla causing it to release adrenaline (A) and noradrenaline.

On the other side the work of Selye (1976) was centered on the role of the adrenal cortex in the stress response. Selye's findings and theories originated the HPA model of stress: during stress adrenocorticotropic hormone (ACTH) is released from cells of the anterior pituitary and elicits secretion of glucocorticoids from the adrenal cortex (Fig. 1).

As already indicated, SA and HPA responses are not independent from each other: in fact both systems are implicated, as indicated by Maickel et al. (1967). Lesions of the sympathetic nervous system or the adrenal, but not the adrenal medulla alone, decrease the ability of the animals to withstand cold while the effect of adrenalectomy can be alleviated by cortisol treatment.

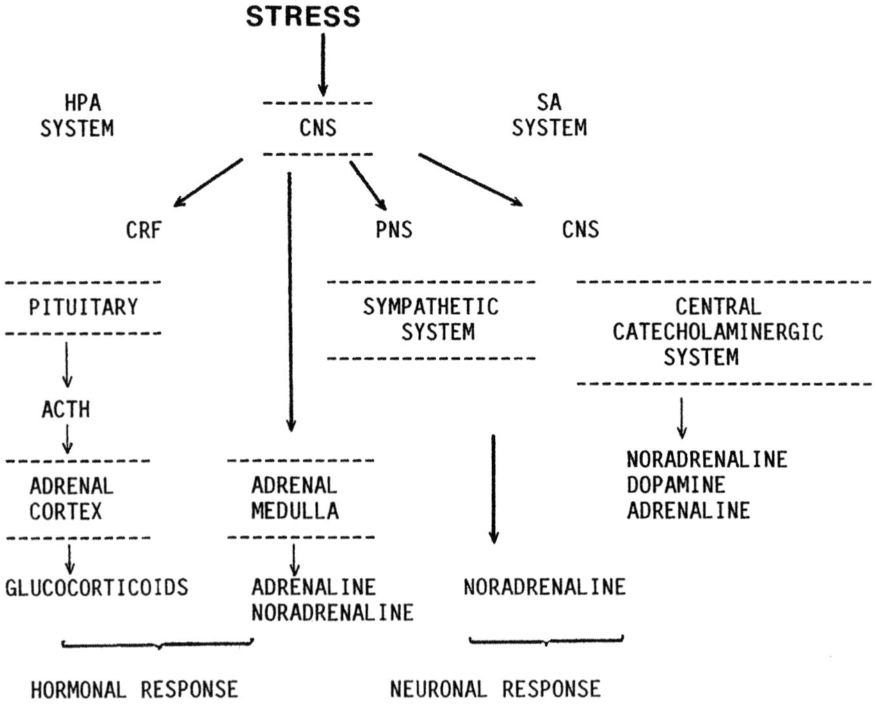

Fig. 1. Hypothalamic-pituitary-adrenocortical (HPA) and sympatho-adrenomedullary (SA) response in stress. The activation of both HPA and SA systems is shown.

The adaptive value of sympathetic activation and corticosteroid secretion is evident in both Cannon's and Selye's conceptions. Adrenaline causes an activation of the glucose supply by accelerating the degradation of glycogen in the liver and by diverting the blood from the viscera and muscles; in addition to that, lipolytic processes are also enhanced. Also glucocorticoids stimulate glycogenolysis and lipolysis, increase plasma glucose, stimulate the conversion of aminoacids to glucose (glyconeogenesis), thus increasing the metabolic supply in order to cope with the stressful situation.

Though the organism copes with stressful events through a number of peripheral reactions, most stress theories do not take into sufficient account the role of the brain, the organ which regulates the various hormonal responses and which is responsible for a number of neural adaptive mechanisms at the central and peripheral level. It is now becoming more and more evident that stress responses are not limited to a concerted reaction of SA and HPA systems at the periphery: they also implicate many neurochemical (monoaminergic) and neuroendocrinological (peptidergic) reactions at the brain level.

1. From the SA side, as indicated by Stone (1975) and Anisman (1978), in addition to the peripheral sympathetic response there is also an activation of cerebral catecholaminergic neurons leading to increased levels of catecholamines in discrete brain regions, increased catecholamine accumulation from tyrosine precursor (Dunn et al., 1981) and enhanced dopamine utilization in brain regions such as the frontal or cingular cortex (Lavielle et al., 1979).

2. From the HPA side it was shown that β-endorphin and β-lipotropin (β-LPH) are released at the brain level along with ACTH during stress (Guillemin et al., 1977): this release is not limited to the hypothalamic level since neurons containing ACTH, β-endorphin and β-LPH exist in other brain regions (Table 1) .

In addition to that, the secretion of a number of other hormones is altered during stress: plasma levels of prolactin and vasopressin are increased, a fact which may be secondary to release of β-endorphin. Also the levels of TSH and growth hormone are enhanced, though the secretion of growth hormone is increased in primates but not in other mammals (Martin et al., 1977).

Table 1. Physiological responses to stress

CNS	Peripheral
Sympathetic-Adrenomedullary (SA)	
Adrenergic System (A ↑)	Sympathetic N.S. (NA ↑, Enkephal. ↑)
Noradrenergic System (NA ↑)	Adrenal Medulla (A ↑, NA ↑, Enkephal. ↑)
Dopaminergic system (DA ↑)	
Serotonergic System (5-HT ↑)	
Hypothalamic-Pituitary-Adrenocortical (HPA)	
Pro-opiomelanocortin (POMC) System	Hypothalamus (CRF ↑)
	↓
	Pituitary (ACTH ↑, β- LPH - β- endorphin ↑)
	↓
	Adrenal cortex (glucorticoids ↑)
	Other Hormones
	* Prolactin ↑
	Growth Hormone ↑ ↓
	TSH-thyroxine ↑
	* Gonadal Hormones ↑
	* β-endorphin mediated

An integrated picture of the brain in relation to stress is emerging from the studies centered on brain peptides and on the relationships existing between peptidergic and aminergic neurons. If we consider the HPA system in light of recent findings we may see (Fig. 2) that the noradrenergic, dopaminergic, serotonergic and cholinergic inputs from the hypothalamus alter the release of corticotropin - releasing factor (CRF), the peptide stimulating the release of ACTH from the pituitary. While centrally acting adrenaline may inhibit CRF-induced ACTH release, dopaminergic neurons seem to exert a stimulating effect on CRF and ACTH release (Mezey et al., 1984).

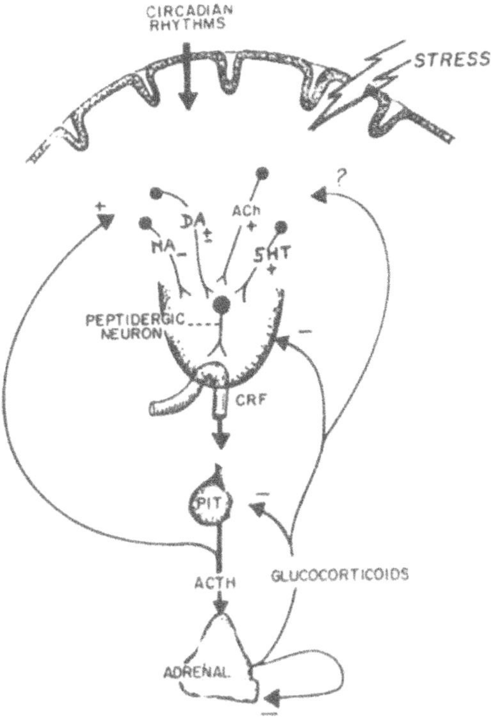

Fig. 2. Noradrenergic, dopaminergic, serotonergic and cholinergic inputs to the hypothalamus alter the release of CRF which stimulates the release of ACTH from the pituitary. ACTH stimulates in turn the release of glucocorticoids from the adrenal cortex. Glucocorticoid and ACTH feedback are also shown.

It must be pointed out that the scheme shown above presents many complications and I will just indicate some critical points:

1. Vasopressin and other peptides may act as a CRF.

2. ACTH is just one of many peptides deriving from the POMC molecule (Fig. 3) which is mostly "cut" into ACTH, β-LPH and β-endorphin in the anterior lobe of the pituitary while α-MSH, CLIP and β-endorphin are produced in the intermediate lobe.

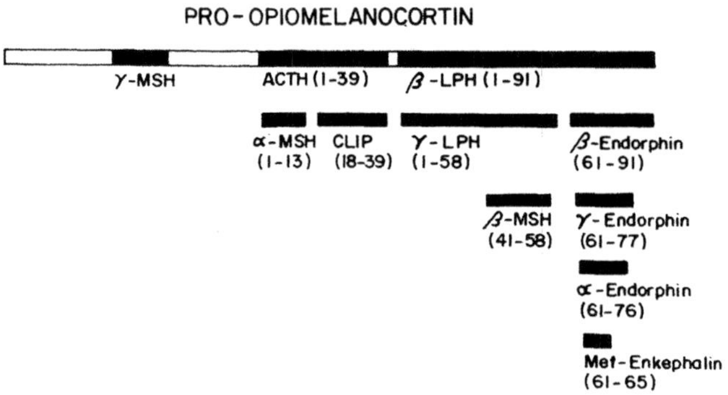

Fig. 3. The pro-opiomelanocortin (POMC) molecule, a precursor of ACTH, MSH, β-endorphin and other peptides. Pituitary anterior lobe cells produce primarily ACTH and β-LPH, while the intermediate lobe and brain cells produce α-MSH, CLIP and β-endorphin.

3. These peptides and their derivatives are active at different receptor sites at both central and peripheral level and produce a number of behavioral modifications ranging from behavioral stereotypies (Gispen and Isaacson, 1981; Cabib et al., 1985), to stress-induced analgesia (Amir et al., 1980) or immunological functions such as chemotaxis (Ruff et al., 1985).

4. There is increasing evidence that peptide hormones may be transported into the brain from the periphery (where they are also synthesized).

5. Experimental data by Rose and Sachar (1981) suggest that CRF modulates, under control of the amygdala and hippocampus, the adrenergic activity.

Following the identification of the tuber-infundibular dopaminergic system (TIDA) and of the role of dopamine in the medio-basal hypothalamus, many studies are now centered on the relationships existing between dopaminergic function and the neuroendocrine response to stress. Prolactin (PRL) and growth hormone (GH) are under the inhibitory control of the TIDA system while endogenous opioids stimulate PRL and GH secretion, thus increasing their plasmatic levels in conditions of stress (Noel et al., 1972). A dopaminergic regulation of CRF, through D1 receptors, takes place at the median eminence of the hypothalamus and the pituitary level (Muller, 1984). Finally, a dopaminergic control of the release of CLIP, α-MSH and β-endorphin has been suggested (Rose and Sachar, 1981). These findings indicate the complex relationships existing at the brain level in terms of noradrenergic and dopaminergic systems and CRF production, as indicated by Figure 4. In general, the hypothesis has been put forward that the dopaminergic system, through the intermediate pituitary lobe, takes part in the response to stress when the adrenergic sympathetic system is overstimulated.

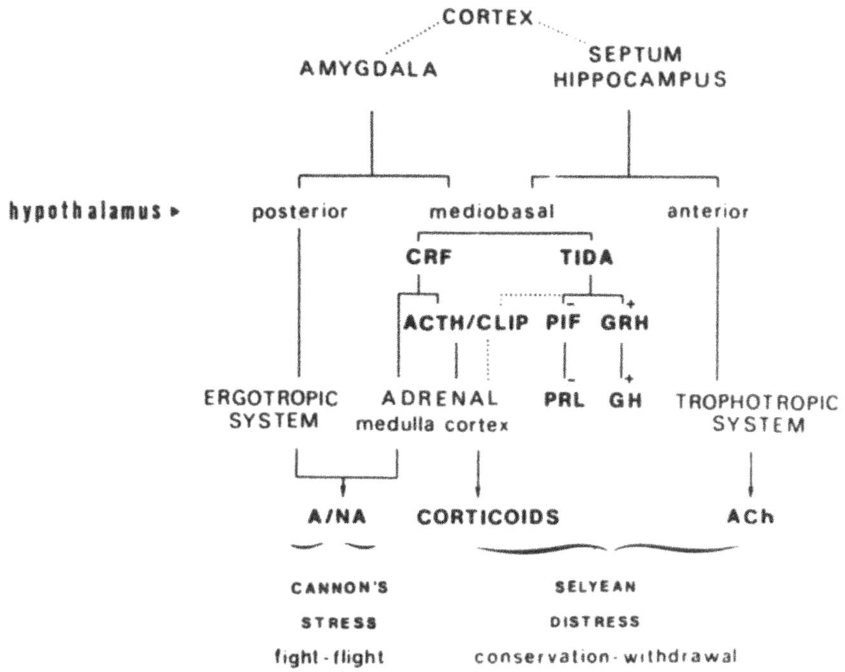

CRF : corticotropic releasing factor
TIDA : tuber-infundibular dopaminergic system
ACTH : adrenocorticotropic hormone
CLIP : corticotropic-like intermediate peptides
PIF : prolactin inhibiting factor
GRH : growth releasing hormone
PRL : prolactin
GH : growth hormone
A : adrenaline
NA : noradrenaline
ACh : acetylcholine

Fig. 4. Neuroendocrine and automatic hypothalamic systems involved in stress reactions. The role of the dopaminergic system is underlined in this figure.

Finally, recent findings (Udelsman et al., 1986) indicate that CRF acts at both central and peripheral level as a key hormone in the integrated response to stress (Table 2).

Table 2. Physiological responses to corticotropin releasing factor (CRF)

CNS	Anterior pituitary	Peripheral
		Adrenal medulla
		↓
Cortex	ACTH ↑	Stimulation of Adenylate Cyclase Activity
Limbic system	Met-enkephalin ↑	↓
		Release of Met-enkephalins, Catecholamines
		↓
		Release of Corticoids ?

CRF is present in the adrenal medulla and sympathetic ganglia: its receptors in the adrenal medulla are coupled to adenylate cyclase activity and can stimulate the secretion of catecholamines and Met-enkephalin.

It has been suggested (Udelsman et al., 1986) that the peptides and catecholamines secreted from the medulla may have an effect on the cortical areas and release glucocorticoids (Fig. 5).

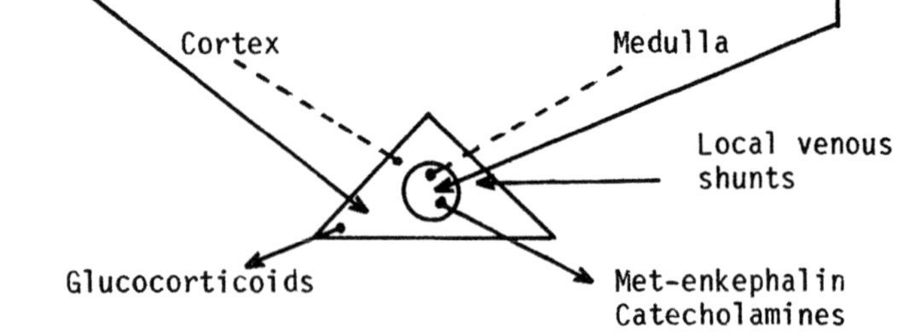

Fig. 5. CRF and integrated stress responses.

If this effect will be definitely established it will be evident that a number of interacting systems take part in stress responses and that CRF has a major role in these processes with actions at all the levels of the neuroendocrine system.

REFERENCES

Amir, S., Brown, Z.W. and Amit, Z. 1980. The role of endorphins in stress: evidence and speculations. Neurosci. Biobehav. Rev., 4, 77-86.

Anisman, H. 1978. Neurochemical changes elicited by stress. In "Psychopharmacology of aversively motivated behavior" (Ed. H. Anisman and G. Bignami). (Plenum Press, New York). pp. 119-172.

Cabib, S., Puglisi-Allegra, S. and Oliverio, A. A genetic analysis of stereotypy in the mouse: dopaminergic plasticity following chronic stress. Behav. Neur. Biol., 44, 239-248.

Cannon, W.B. and de la Paz, D. 1911. Emotional stimulation of adrenal secretion. Am. J. Physiol., 28, 64-70.

Dunn, A.J. and Kramarcy, N.R. 1984. Neurochemical responses in stress: relationships between the hypothalamic - pituitary - adrenal and catecholamine systems. In "Handbook of Psychopharmacology" (Ed. L.L. Iversen, S.D. Iversen and S.H. Snijder). (Plenum Press, New York). pp. 455-515.

Dunn, A.J., Childers, S.R., Kramarcy, N.R. and Villiger, J.W. 1981. ACTH-induced grooming involves high affinity opiate receptors. Behav. Neural. Biol., 31. 105-109.

Gispen, W.H. and Isaacson, R.L. 1981. ACTH-induced excessive grooming in the rat. Pharmacol. Ther., 12, 209-246.

Guillemin, R., Vargo, T., Roissier, J., Minick, S., Ling, N., Rivier, C., Vale, W. and Bloom, F. 1977. β-endorphin and adrenocorticotropin are secreted concomitantly by the pituitary gland. Science, 197, 1367-1369.

Hitchings, F.W., Sloan, H.J. and Austin, J.B. 1913. Laboratory studies of the activity of the brain and the adrenal in response to specific stimuli. Cleveland Med. J., 12, 684-691.

Lavielle, S., Tassin, J.P., Thierry, A.M., Blanc, G., Herve, D., Barthelemy, C. and Glowinski, J. 1979. Blockade by benzodiazepines of the selective high increase in dopamine turnover induced by stress in mesocortical dopaminergic neurons of the rat. Brain Res., 168, 585-594.

Maickel, R.P., Matussek, N., Stern, D.N. and Brodie, B.B. 1967. The sympathetic nervous system as a homeostatic mechanism. J. Pharmacol. Exp. Ther., 157, 103-110.

Mason, J.W. 1968. A review of psychoendocrine research on the sympathetic-adrenal medullary system. Psychosomat. Med., 30, 631-653.

Martin, J.B., Reichlin, S. and Brown, G.M. 1977. Clinical neuroendocrinology. (F.A. Davis, Philadelphia).

Mezey, E., Kiss, J.Z., Skirboll, L.R., Goldstein, M. and Axelrod, J. 1984. Increase of corticotropin-releasing factor staining in rat paraventricular nucleus neurones by depletion of hypothalamic adrenaline. Nature, 310, 140-141.

Muller, E.E. 1984. Neuroendocrinology. (EMSI, Rome).

Noel, G.L., Shu, H.K., Stone, G. and Frantz, A.G. 1972. Human prolactin and growth hormone release during surgery and other conditions of stress. J. Cli. Endocrinol., 35, 840-851.

Rose, R.M. and Sachar, E. 1981. Psychoneuroendocrinology. In "Textbook of endocrinology" (Ed. R.H. Williams). (Saunders, Philadelphia). pp. 646-671.

Ruff, M.R., Pert, C.B., Weber, R.J., Wahl, L.M., Wahl, S.M. and Paul, S.M. 1985. Benzodiazepine receptor-mediated chemotaxis of human monocytes. Science, 229, 1281-1283.

Seleye, H. 1976. Stress in health and disease (Butterworths, Boston).

Stone, E.A. 1975. Stress and catecholamines. In "Catecholamines and behavior, Vol. 2". (Ed. A.J. Friedhoff). (Plenum Press, New York). pp. 31-72.

Udelsman, R.,Harwood, J.P., Millan, M.A., Chrousos, G.P., Goldstein, D.S. Zimlichan, R., Catt, K.J. and Aguilera, G. 1986. Functional corticotropin releasing factor receptors in the primate peripheral sympathetic nervous system. Nature, 319, 147-150.

ENDOCRINE ASPECTS OF STRESS: EVALUATION OF STRESS REACTIONS IN FARM ANIMALS

J. Ladewig

Institut für Tierzucht und Tierverhalten - FAL
Trenthorst, 2061 Westerau, BRD

ABSTRACT

Any evaluation of stress reactions must include a consideration of the kind of stress stimulus being studied as well as the kind of animal being subjected to the stress stimulus. For farm animals subjected chronically to aversive situations that cannot be alleviated through some behavioral response, an activation of primarily the hypothalamo-pituitary corticoadrenal system should be expected. Since this system is activated not only by stress stimuli, however, but is also influenced by endogenous factors giving rise to circadian and ultradian variations, these stress independent aspects of corticoadrenal activity must be determined, before any stress dependent changes can be revealed. Similarly, any evaluation of the effect of stress on an animal must include the same consideration of the type of stress stimulus and the kind of animal being studied. In animal husbandry, the effect of stress on growth, reproduction and resistance to disease is of particular interest. The possible interaction between stress reactions and the endocrine aspects of these other body functions is obvious.When studying this interaction, however, it is necessary to consider more carefully than has been done so far the biological rhythms of the hormone secretion, not only as far as the stress hormones are concerned, but also as far as other hormones are concerned.

INTRODUCTION

It is my task to discuss how the fundamental aspects of stress and the endocrine system, as presented in the previous chapter, can be applied to stress research in farm animals. Particularly two questions are of importance, firstly, which hormonal changes occur during stress and, secondly, how do these changes affect the animals.

Before adressing these specific questions, however, it is necessary to analyse in some detail not only the stimuli that lead to a stress response, but also the kind of animal subjected to the stimuli, since these factors have an effect on the response.

As demonstrated by Selye half a century ago, an animal

that is subjected to a stress stimulus reacts with some kind of stress response (Selye, 1956). Selye called this response nonspecific because, according to his observations, a wide spectrum of stressors resulted in the same kind of stress reactions: Adrenal hypertrophy, lymphatic hypotrophy, and gastrointestinal ulceration. Although we know now that all three reactions are not always present following presentation of all stressors, the present concept of stress still encompasses "all possible extra-individual events capable of evoking a broad spectrum of intra-individual responses mediated by a complex filter labeled "individual differences"" (Veith-Flanigan and Sandman, 1985).

STRESS STIMULI

These extra-individual events or stress stimuli can be described in a qualitative as well as a quantitative way. Qualitatively, it is possible to describe a stressor according to its physical properties as a thermal, chemical, electrical stressor, etc. Physical stimuli alone, however, may not induce a stress reaction. In a series of studies, Mason and coworkers (Mason et al., 1974), found that stress stimuli only lead to a stress reaction if the animal also shows an emotional response. For instance, the corticosteroid concentration in the urine of rhesus monkeys increased when room temperature was increased suddenly, whereas a gradual temperature increase of the same magnitude was followed by a decrease in adrenal activity. Apparently, the suddenness of the temperature change produces an emotional reaction which leads to the adrenal activation.

A similar result has since been found in other studies. Suzuki (1983) showed that the cortisol concentration of blood collected from dogs engaged in physical activity increased markedly only, when a stage of physical exhaustion was reached. Dogs that were still fit after the exercise, and presumably less emotionally upset, showed only a slight increase in adrenal activity.

Based on these and similar studies, Mason (1971)

concluded that various stressors exert their effects most potently when the emotional status of an organism is involved, and that the stress reaction is not a nonspecific reaction to a great variety of stimuli, as Selye suggested, but rather a specific reaction to psychological stressors. Although not all physiological stress reactions are mediated by emotions, Masons conclusion remains valid that "psychological stimuli are among the most potent of all stimuli to affect the pituitary adrenal cortical system" (Mason, 1971).

As regards the quantity of a stress stimulus, two aspects must be considered: Intensity and duration. The physical intensity of some stress stimuli can be measured at their source, but because of individual differences in sensitivity to, and perception of the stimuli, their physiological impact is difficult to assess. In comparison, measurement of the duration of a stress stimulus may, at first hand, seem simpler. Thus, it is usual to distinguish between acute and chronic stressors. Transportation for a few hours may be an example of an acute stressor and confinement over several months an example of a chronic stressor. A distinction between acute and chronic stress is, however, not sufficient to describe all stress situations. A regular or irregular repetition of an acute stressor over a longer period constitutes a stress situation which has been termed chronic intermittent stress (Burchfield, 1979). Unfortunately, most investigators do not distinguish between chronic and chronic intermittent stressors, despite the fact that the stress response may be different to these states. If an animal is repeatedly subjected to an acute stressor, the corticoadrenal response changes. For stressors of low intensity, the stress response may even disappear altogether. Following repeated exposure of a more intense stressor, the maximal stress response may occur prior to the stimulus onset and may even terminate before actual onset. Furthermore, this anticipatory stress response is smaller than the initial reaction. The reduction is not due to adrenal exhaustion, since presentation of a novel acute stressor or ACTH injection produce a stress response of original size (Ramade and Bayle, 1984).

When comparing chronic and chronic intermittent stress it is, in fact, questionable whether a true chronic stress situation ever exists. It is known, for instance, that cattle kept on slatted floor are subjected to various aversive stimuli. It is possible, however, that these negative stimuli are more intense when an animal is walking on the slatted floor than when the animal is lying down on it (or vice versa). Furthermore, as regards the emotional involvement in the mediation of a stress stimulus, it is possible that an animal experiences a constant stress stimulus differently at different times. A tethered sow, for instance, may experience confinement more intensely during periods of activity than during periods of preoccupation such as eating or during periods of inactivity such as sleeping. Therefore, although these environmental factors are constantly present and could be termed chronic stressors, it may be that their effect on the individual vary in intensity. These factors could therefore be considered as chronic intermittent stressors.

INDIVIDUAL DIFFERENCES

The stress stimulus is mediated and modified by the individual. Therefore, the stress response depends on the kind of animal being examined. Animal factors such as species and breed differences and differences between various family lines may influence the stress reaction. Male-female differences may originate from the CNS or from differences in sex hormone secretion. Variability due to age is probably less pronounced, but a reduced corticoadrenal capacity in older animals has been reported (Riegle and Nellor, 1967). One area, the importance of which so far has not been sufficiently acknowledged, is the influence of early experience. Studies on the effect of early handling on behavior and physiological responses later in life (Levine et al., 1967), indicate that a critical or sensitive period may exist, during which emotionality is modified and corticoadrenal response reduced.

In work with farm animals, individual differences in both behavioral and physiological responses are often so large that apparent stress induced changes may be masked by individual differences. In an attempt to analyse this problem systematically, a study was undertaken in which the cortico-adrenal reactivity of castrated male pigs was determined at the age of 4.5 months. The animals were anesthetised and the plasma cortisol concentration 3 hours later was determined. The subjects were then designated according to their cortisol levels as high reacting, middle high, middle low, and low reacting. Seven weeks later half of the animals in each group was tethered in individual pens with partially slatted floors without straw and the other half was kept as controls in groups on straw. Six weeks later at the age of 8 months, the animals underwent an adrenal function test: each animal was given 200 IU ACTH intravenously (Synacthen, Ciba-Geigy) and the resulting cortisol increase was determined in blood samples collected over a 5-h period. Within each reaction group, tethered animals showed a higher cortisol increase than control animals. However, this difference was considerably smaller than the variation between groups. Therefore, had individual responsiveness not been taken into account, the effect of the environment on adrenal function could not have been so clearly demonstrated (von Borell and Ladewig, 1986).

STRESS RESPONSE

After having considered the stimulus side of the stress situation and the influence of the individual, we can finally focus on the stress reaction. Two endocrine systems are of particular interest: The sympathetic adrenal medullary system and the hypothalamo-pituitary adreno-cortical system. Despite the interconnections between the two systems described in the previous chapter, the activation of them varies according to the stressor experienced, as reviewed in detail by Henry and Stephens (1977). Briefly summarized, if the stressor constitutes a threat to the animal, it reacts with a fight or flight response. The amygdala and sympathetic adrenal-medullary system are activated with a resulting increase in catecholamine secretion. In this situation the animal is still able to alter the situation through some behavioral response. If, on the other hand, the animal has lost control over the situation, or if it is no longer able to cope with the stressor, behavioral activity is suppressed. This effect is mediated by activation of the hippocampal-septal part of the limbic system which also activates the hypothalamo-pituitary adrenal axis.

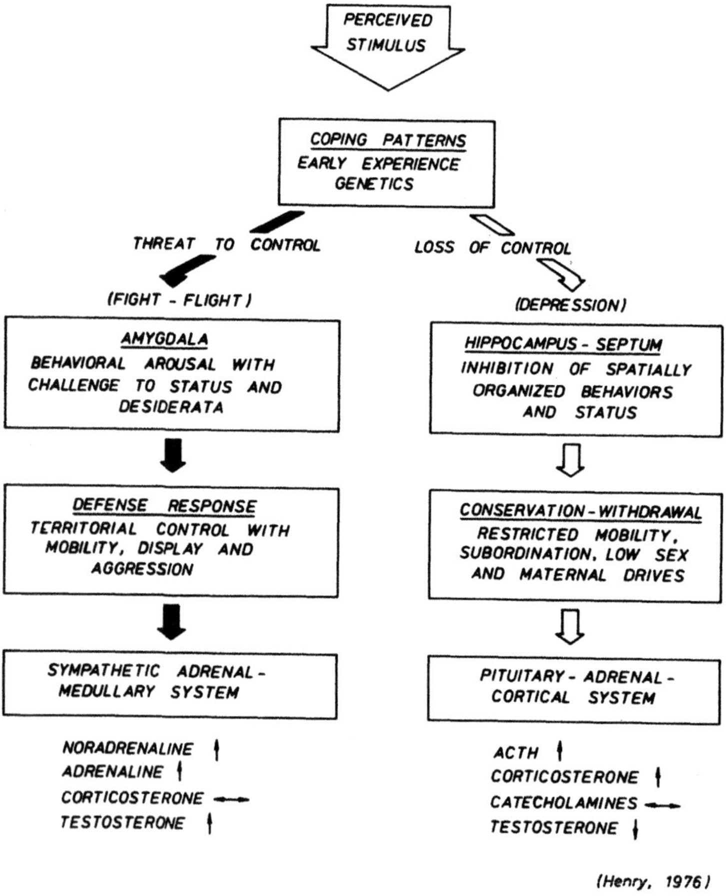

A growing body of literature supports this possible involvement of coping mechanisms. For instance, as first demonstrated by Weiss (1972), rats subjected to electroshocks showed a lower percentage of gastric ulceration, if they were able to exert some control over the shock administration as compared to yoked rats that did not have this possibility. More recently, Schuurman (cited by Wiepkema, 1985) demonstrated that fighting in male rats caused increased corticoadrenal activity, but that the increase was more enhanced in the losing animal than in the winning animal; this difference was established before the fight was terminated. Apparently, the outcome of the fight was known relatively early, and the physiological mechanisms adjusted accordingly.

It was mentioned earlier that the corticoadrenal response decreases following repeated exposure to a stressor and that the rise in cortisol levels may begin before the onset of the stressor. In a series of experiments on rats, Kvetnansky and coworkers demonstrated that an opposite change occurs in the sympathetic adrenal medullary system. Thus, during the first exposure to immobilization stress, adrenaline and nor-adrenaline secretion is relatively low compared to later exposures. During the initial immobilization, the subjects have not learned the consequences of the situation and respond primarily with increased corticosterone secretion. During later exposures, the animals know the outcome of the stress situation, are therefore presumably able to cope with it, and react predominantly through adrenaline-noradrenaline secretion (Kvetnansky and Mikulaj, 1970; Kvetnansky et al., 1970; Mikulaj and Mitro, 1973).

In modern agriculture an important question is whether intensive husbandry systems are too stressful for the animals and whether animals kept in these systems are suffering. Since the animals are kept in these systems over longer periods, the stressors are chronic or chronic intermittent rather than acute. Furthermore, since the animals usually show no fight or flight type of reaction, an activation of primarily the hypothalamo-pituitary adrenal axis should be expected.

Although many studies have attempted to demonstrate an increased corticoadrenal activity in intensively kept farm animals, the results are far from clear. For example, Barnett et al. (1981) found that gilts confined individually over 2 weeks showed similar corticosteroid levels in the morning as gilts kept in groups, whereas the levels in the afternoon were higher in the individually penned animals. Becker et al. (1985) found higher morning cortisol levels in gilts kept tethered over 3 weeks in comparison with individually penned gilts, whereas afternoon levels were similar in the two groups. Thus, in one study the supposedly stressed gilts showed increased adrenal activity in the afternoon but not in the morning, and in another study supposedly stressed gilts showed increased adrenal activity in the morning but not in the afternoon.

In the following some reasons are briefly described why these kind of studies can give such divergent results. Firstly, since the secretion of corticosteroids shows not only diurnal but also ultradian variations (e.g. Hellman et al., 1970; Thun et al., 1981), it is necessary to consider these "stress independent" or endogenous variations, before any "stress dependent" changes can be recognized. In most diurnal mammals, episodes of adrenal activity occur primarily between early morning and early afternoon. The interval between secretory episodes is approximately 90 min, and the total frequency of episodes per 24 h is about 10. Obviously, the contribution of these endogenous variations in adrenal activity must be analysed before the effects of stressors can be identified (Ladewig, 1984).

Secondly, whereas the adrenal response to an acute stressor is usually synchronised between animals through the onset of the stressor, such a synchronisation is most likely absent in animals subjected to chronic or chronic intermittent stress. Therefore, an increased adrenal response to this type of stressor can only be revealed when the frequency and amplitude of the secretory episodes are measured over an extended period.

In a study on bulls kept tethered on partially slatted floor, it was found that, two and three days after tethering, the average amplitude of secretory episodes was increased compared to before tethering, whereas the frequency of secretory episodes did not change. In between these elevated episodes, cortisol concentrations were similar to those of the control situation (Ladewig, 1984). In other words, the stress response to this chronic stressor was only seen during short bursts and occurred at different times in different animals.

Thirdly, the plasma hormone concentration is not only dependent upon hormone secretion, but also upon removal of the hormone from the blood stream. Although studies in humans have indicated that the metabolic clearance rate remains relatively constant, possible changes in this factor must also be considered to get a true picture of adrenal function (De Lacerda et al., 1973). Moreover, a proportion of the circulating corticosteroids is bound to proteins in the blood, particularly to transcortin, and only the unbound hormone possesses biological activity. Therefore, since the effects of stress on the unbound and bound portions may differ, the degree of hormone binding should also be determined, as suggested by Barnett et al. (1981).

In conclusion, if the corticoadrenal activity is to be used as a stress indicator, all these various possible confounding factors must be considered. As an alternative, an indirect determination of corticoadrenal activity may be employed, such as determination of corticosteroids in urine or the adrenal function test mentioned earlier. A determination of the plasma corticosteroid concentration at one or a few points of time alone is not sufficient to reveal a stress induced change in adrenal activity.

THE EFFECT OF STRESS

The second question that needs to be adressed is: how does the stress response affect the animal. The answer to this question is, without doubt, much more complex than any of the answers we have been looking for so far. Since it is also the theme of the next topics of this meeting, only a brief introduction will be given.

As a starting point, it is important to keep in mind the following fundamental statement: the stress response increases the resistance to stress (Munck et al., 1984). That is, in the acute stage, the stress response is, in general, beneficial to the individual. On the other hand, if the stress response continues for too long, it may turn harmful to the individual. Therefore, when we investigate the possible effects of stress or when we try to use the stress response to interpret the welfare status of an animal, it is necessary to consider the same factors as mentioned earlier, such as whether a stressor is acute or chronic and what kind of animal is being examined.

Apart from the welfare question, particularly three areas that may be negatively influenced by stress, are of interest in farm animal production: growth, reproduction, and resistance to disease. The possible hormonal involvement in the disturbance of these body functions is obvious, at least as far as growth and reproduction is concerned, both being regulated by pituitary hormones. However, when investigating the possible connections between stress mechanisms and these other functions, it is necessary first to establish at which level such connections may exist. One possibility is that those central nervous mechanisms that are responsible for the stress response simultaneously affect these other body functions, that is, that interconnection exists at the central level. Another possibility is that the influence on growth, reproduction and the immune system are secondary to the stress responses. For instance, a possible interconnection between the hypothalamo-pituitary adrenal axis and the hypothalamo-pituitary gonadal axis may involve ACTH which, theoretically, can interact either on the central level, on the pituitary level, or on the gonadal level. The interconnection may, of course, also involve the corticosteroids or CRF at any of the levels mentioned.

Secondly, whatever the connections are, it is necessary to remember that the secretion of all pituitary hormones is episodic, so that any study on the effect of stress on LH, FSH, GH, etc. must analyse the changes in the episodic secretion of these hormones (i.e. changes in amplitude or frequency of secretory episodes). I.e., also in this kind of studies blood samples must be collected frequently over an extended period.

A number of studies have examined the effect of ACTH or corticosteroids on some aspect of reproduction (e.g. Fuquay and Moberg, 1983; Liptrap and McNally, 1976; Matteri et al., 1984). A common method is to treat an animal with synthetic corticoids or ACTH which yield elevated plasma concentrations for an extended period. Since these treatments are often nonphysiologic not only as far as dosage is concerned, but especially with respect to biological rhythm, the results from such studies must be interpreted with caution. What is needed in this type of studies is a hormone treatment that resembles the stress induced hormone secretion more closely. For instance, rather than injecting a depot form of ACTH intramuscularly, the hormone should be injected intravenously in physiological dosages in a pulsatile manner, in order to simulate the secretion during stress.

In conclusion, when discussing stress and, in particular, when studying stress, it is necessary to take a much more detailed look at this phenomenon then has been done so far. It is important to realize that what happens in a stress situation is so complex that it cannot be described with just one word. And so long as we use expressions such as "stress and the endocrine system", "stress and the immune system", "stress and behavior", we wrongly propagate the impression that stress is a well-defined, straight forward process.

ACKNOWLEDGEMENT

Part of the research described in this report was funded by the German Research Council. Synacthen was generously donated by Drs. R. Andreatta and K. Scheibli, Ciba-Geigy.

REFERENCES

Barnett, J.L., Cronin, G.M. and Winfield, C.G. 1981. The effects of individual and group penning of pigs on total and free plasma corticosteroids and the maximum corticosteroid binding capacity. Gen. Comp. Endocrinol., 44, 219-225.

Becker, B.A., Ford, J.J., Christenson, R.K., Manak, R.C., Hahn, G.L. and DeShazer, J.A. 1985. Cortisol response of gilts in tether stalls. J. Anim. Sci., 60, 264-270.

von Borell, E. and Ladewig, J. 1986. Möglichkeiten der Erfassung von chronischen Belastungsreaktionen beim Schwein mit Hilfe von Nebennierenrinden-Funktionsprüfungen und ethologischen Merkmalen. In "Aktuelle Arbeiten zur artgemäßen Tierhaltung", KTBL-Schrift, 311, 30-36.

Burchfield, S.R. 1979. The stress response: A new perspective. Psychosom. Med., 41, 661-672.

DeLacerda, L., Kowarski, A. and Migeon, D.J. 1973. Diurnal variation of the metabolic clearance rate of cortisol. Effect on measurement of cortisol production rate. J. Clin. Endocrinol. Metab., 36, 1043-1049.

Fuquay, J.W. and Moberg, G.P. 1983. Influence of the pituitary-adrenal axis on the induced release of luteinizing hormone in rams. J. Endocr., 99, 151-155.

Hellman, L., Nakada, J., Curti, J., Weitzman, E.D., Kream, J., Roffwarg, H., Ellman, S., Fukushima, D.K. and Gallagher, T.F. 1970. Cortisol is secreted episodically by normal man. J. Clin. Endocrinol., 30, 411-422.

Henry, J.P. 1976. Mechanisms of psychosomatic disease in animals. Adv. Vet. Sci. Comp. Med., 20, 115-145.

Henry, J.P. and Stephens, P.M. 1977. Stress, health and the social environment. A sociobiologic approach to medicine. (Springer, New York). pp. 118-140.

Kvetnansky, R. and Mikulaj, L. 1970. Adrenal and urinary catecholamines in rats during adaptation to repeated immobilization stress. Endocrinology, 87, 738-743.

Kvetnansky, R., Weise, V.K. and Kopin, I.J. 1970. Elevation of adrenal tyrosine hydroxylase and phenylethanolamine-N-methyl transferase by repeated immobilization of rats. Endocrinology, 87, 744-749.

Ladewig, J. 1984. The effect of behavioral stress on the episodic release and circadian variation of cortisol in bulls. In "Proc. Internat. Congr. Appl. Ethol. Farm. Anim." (Ed. J. Unshelm, G. van Putten and K. Zeeb). (KTBL, Darmstadt). pp. 339-342.

Levine, S., Haltmeyer, G.C., Karas, G. and Denenberg, V. 1967. Physiological and behavioral effects of infantile stimulation. Physiol. Behav. 2, 55.

Liptrap, R.M. and McNally, P.J. 1976. Steroid concentration in cows with corticotropin-induced cystic ovarian follicles and the effect of prostaglandin F2α and indomethacin given by intrauterine injection. Am. J. Vet. Res., 37, 369-375.

Mason, J.W. 1971. A re-evaluation of the concept of "non-specificity" in stress theory. J. Psychiat. Res., 8, 323-333.

Mason, J.W. 1984. Specificity in the organization of neuro-endocrine response profiles. In "Frontiers in Neurology and Neuroscience Research" (Ed. P. Seeman and G. Brown). (Univ. of Toronto). pp. 68-80.

Matteri, R.L., Watson, J.G. and Moberg, G.P. 1974. Stress or acute adrenocorticotrophin treatment suppresses LHRH-induced LH release in the ram. J. Reprod. Fert., 72, 385-393.

Mikulaj, L. and Mitro, A. 1973. Endocrine functions during adaptation to stress. Adv. Exp. Med. Biol., 33, 631-638.

Munck, A., Guyre, P.M. and Holbrook, N.J. 1984. Physiological functions of glucocorticoids in stress and their relation to pharmacological actions. Endocr. Rev., 5, 25-44.

Ramade, F. and Baylé, J.D. 1984. Adaptation of the adreno-cortical response during repeated stress in thalamic pigeons. Neuroendocrinology, 39, 245-250.

Riegle, G.D. and Nellor, J.E. 1967. Changes in adrenocortical function during aging in cattle. J. Gerontol., 22, 83-86.

Schuurman, T. Cited by Wiepkema, P.R. 1985. Biology of fear. In "Second Eur. Symp. Poultry Welfare" (Ed. R.M. Wegener). (FAL, Braunschweig). pp. 84-92.

Selye, H. 1956. The Stress of Life. (McGraw-Hill, New York).

Suzuki, T. 1983. Physiology of adrenocortical secretion. Frontiers of Hormone Research, 11. (Karger, Basel).

Thun, R., Eggenberger, E., Zerobin, K., Luscher, T. and Vetter, W. 1981. Twentyfour-hour secretory pattern of cortisol in the bull: Evidence of episodic secretion and circadian rhythm. Endocrinol., 109, 2208-2212.

Veith-Flanigan, J. and Sandman, C.A. 1985. Neuroendocrine relationships with stress. In "Stress. Psychological and Physiological Interactions" (Ed. S.R. Burchfield). (Hemisphere, Washington). pp. 129-161.

Weiss, J.M. 1971. Effects of coping behavior with and without a feedback signal on stress pathology in rats. J. comp. physiol. Psychol., 77, 22-30.

INTRODUCTION

In the field of immunology astonishing progress is observed. Not only because of the development of highly sophisticated measuring techniques that led to the discovery of many unexpected details about the finer mechanisms of immune processes, but also because of the finding that this same immune system does not operate independently from what goes on in the neural system. Since stress always appears to be associated with some form of neurobiological state of uncertainty - as elaborated in later chapters -, the link between stress and immunological capabilities is obvious. The relationship between stress and the immune system is described by Ballieux and Heijnen, who show or refer to most amazing data. For instance, the social position of a rat living in a colony significantly determines immunological characteristics of the same rat (cf. the contribution of Bohus et al.). Further, although severe and acute stress may suppress the immune response of the animal involved, moderate forms of stress may do the reverse and enhance such a response.

Of course, an important question is then whether or not stress (mostly) reduces the immune response in a rather unspecific way (e.g., by means of corticosteroids that enter lymofyctes as shown by Siegel in the second contribution), or by regulating immune mechanisms by direct autonomous innervation of organs like the thymus, spleen and others (cf. J. Immunology, Suppl., Vol. 135, Aug. 1985). This question is far from solved, as indicated by Ballieux and Heijnen in their discussion of conditioning of the immune response. Irrespective of what the final answer might be, the essential point is that the immune system in one way or another listens to what goes on in the brain. On the other hand, it is now known that also brain activity on the hypothalamic level reflects what goes on in the immune system, as shown by Besedovski (cited by Ballieux and Heijnen).

In this matter farm animals are no exception, as described by Siegel. This state of affairs has great implications for our thinking about farm animal health. If we keep in mind that stress not only results from so called physical stressors (inflicting some somatic damage), but also from so called psychological stressors (inhibiting the performance of essential behaviour programs for instance), a large array of environmental conditions may interfere with the organism's health. It is a pity that until now we have no good picture how far stress induced immunosuppression influences later immune responses to the same pathogen.

The interaction between the nervous and the immune system is a most fascinating one. It is based on the production and release of a number of compounds, among which the neuropeptides are again conspicuous. This point is touched on in the research of Siegel, who showed the existence of ACTH-like compounds released by lymphocytes. For the moment there seems to be no end to the amazing story of neuropeptides modulating so many highly relevant processes.

These two chapters emphasize that really new insights are not far ahead. However, progress can only be made in so far we succeed in integrating the bewildering mass of biological data now emerging.

STRESS AND THE IMMUNE SYSTEM.

R.E. Ballieux[*] and C.J. Heijnen[**]

[*]Department of Clinical Immunology
University Hospital, 3511 GV Utrecht, The Netherlands
[**]Department of Pediatric Immunology,
University Hospital for Children and Youth
"Het Wilhelmina Kinderziekenhuis", 3512 LK Utrecht, The Netherlands

ABSTRACT

The brain and the nervous system can influence immune responses along two pathways. One of these is the network of the nervous system itself, and the other one involves the hypothalamus/pituitary/adrenal axis and corresponding (peptide) hormones. Both cell mediated as well as antibody dependent immune functions can be modulated following confrontation of the organism with stressful environmental conditions. In this paper the effects of emotional stimuli on the immune system are discussed. Attention is given to the finding that immune responses can be conditioned. Recent data on a possible feedback loop from immune cells to the brain are discussed in relation to the observation that lymphocytes can produce neuroendocrine-like peptide hormones.

INTRODUCTION

There is solid evidence that the nervous system can influence the immune response (Ader, 1981). Consequently, the mental state of an individual may determine the activity of the defence system. Indeed, psychosocial factors, including stressful conditions have been reported to modulate immune function. A number of investigators have shown that when experimental animals are subjected to (acute) stress, the body's defence against micro-organisms and tumor cells is impaired (Monjan, 1981). In most animal experiments the stressors applied are often of physical nature, such as exposure to repeated electric footshocks or subjection to rotation, immobilization or cold temperature. It can be argued that these stressful conditions do not necessarily reflect daily life situation. In this respect it is interesting to note that in recent years the number of studies carried out in animals using stress stimuli of psychological nature, are steadily increasing. In the framework of these investigations it became apparent that the interaction of the brain, perceiving the (stressful) environmental stimulus, and the immune system can follow two pathways (Fig. 1). One is represented by the network of the nervous system, allowing direct innervation of lymphoid tissues. The second one involves the hypothalamus/pituitary/adrenal axis and corresponding hormones as well as neuropeptides released by the brain (Oliverio, 1987).

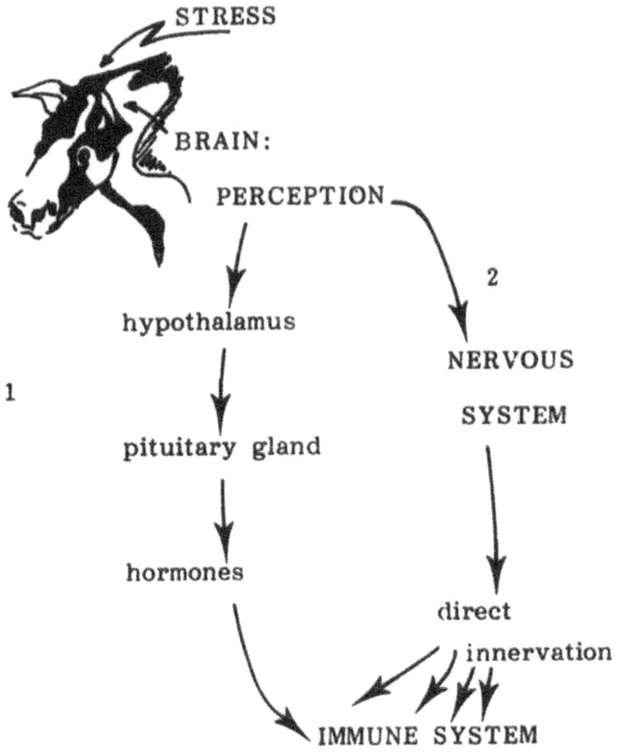

Fig. 1 - Two pathways (1: neuroendocrine and 2: nervous system) linking the brain and the immune system.

This paper will not cover the entire area of neuroimmunomodulation since in the last few years several reviews dealing with this subject have been published (Ballieux, 1984a; Goetzl, 1985; Guillemin et al., 1985; Plotnikoff et al., 1986). It is the aim of the present contribution to focus on a few interesting aspects of the interaction between the nervous system and the immune system.

IMMUNOLOGICAL PARAMETERS IN STRESS RESEARCH

The workhorses of the immune system are the lymphocytes. These cells develop from a stem cell which is localized in the bone marrow to give rise to T lymphocytes and B lymphocytes. The latter lineage comes to full maturation in the bone marrow itself (hence they are called B cells), whereas immature T cells leave this site to spend a short period of their life in the thymus to become fully differentiated (thymus-derived) T lymphocytes. B cells carry surface immunoglobulins which serve as the receptors for antigen; they are precursors of antibody forming cells. Interaction of antigen with the corresponding receptor on B cells in general causes this cell to transform into plasmacells which produce and secrete specific antibodies (Fig. 2).

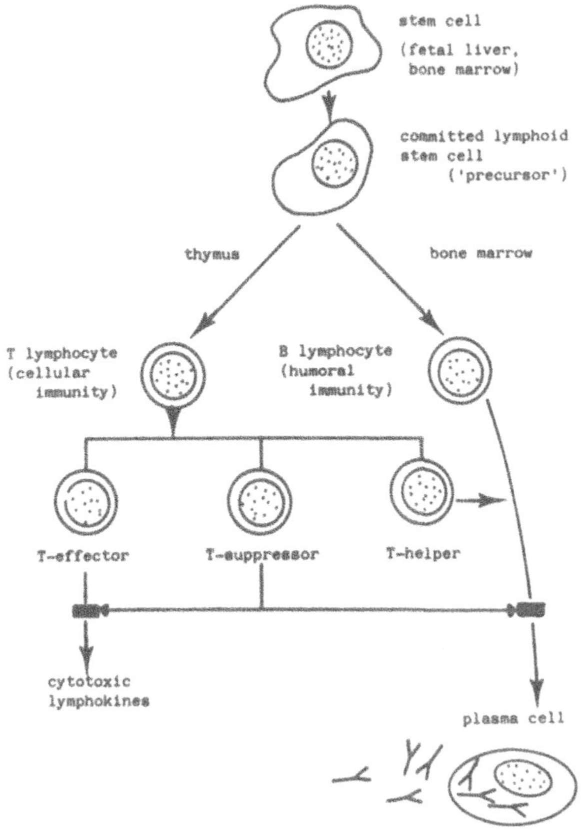

Fig. 2 - Schematic representation of the development of the cells of the immune system (T and B lymphocytes).

T lymphocytes can be distinguished functionally into effector T cells and regulatory T cells. Effector T cells include those that are able by cytotoxic action to directly kill target cells (e.g. tumor cells or virus-infected cells) or that cause delayed-type hypersensitivity reactions by producing a variety of lymphoid factors (lymphokines) which mediate pharmacological reactions. Regulatory T cells either enhance T cell mediated immune reactivity or antibody production (T helper cells) or diminish these immune reactivities (T suppressor cells). In stress related analysis of immune function often use is made of the property of

T and B cells to proliferate in culture systems after stimulation with mitogens like phytohaemagglutinin (PHA), concanavalin A (ConA) or lipopolysaccharide (LPS). In vivo studies frequently include antibody formation after challenge with an antigen, or T cell mediated functions (e.g. delayed hypersensitivity and graft rejection).

It should be emphasized that the use of these read-out systems, which are of relatively simple nature, in most cases does not yield insight in the mechanisms which lead to stress-induced alterations in immune reactivity (Ballieux, 1984b). The outcome of e.g. a PHA-induced lymphocyte response in vitro is the final result of a rather complicated, cooperative type of interaction of non-lymphoid mononuclear white blood cells (monocytes) and different types of T lymphocytes. In case of a diminished response in a stressful condition, the monocytes may have failed to produce sufficient growth factor for the T lymphoyctes or may have produced a suppressive amount of prostaglandins. The T cells may have changed in sensitivity to T cell growth factors produced by monocytes or by T cells themselves. Alternatively, stress may change the balance between the various regulatory T cell subsets, due to an altered circulation pattern which may result in a change in composition of cells in the lymphoid organs. Furthermore, the PHA-induced responsiveness may differ when measured in (defibrinated) whole blood cultures or in cultures of isolated blood lymphocytes in control plasma. In the former case neuroendocrine hormones released in the bloodstream by stress may influence immune reactivity (Johnson et al., 1982; Heijnen et al. 1986a; Heijnen et al., 1986b).

However, although in many paradigms used only an overall effect is measured, it is clear from a great number of solid studies that immune reactivity can be modulated by the central nervous system.

THE EFFECT OF EMOTIONAL STIMULI ON IMMUNE FUNCTION

Emotional stimuli of psychosocial nature have mainly been studied in relation to social hierarchy in rat colonies or using the resident-intruder paradigm. It was found that in the colonies the number of lymphocytes in the peripheral blood of the rats significantly correlated with escape- and flight behaviour. The proliferative response of T lymphocytes after stimulation in vitro with the mitogen ConA also showed a correlation (r=0.63) with social status in the colony (Koolhaas et al., 1986; Ballieux et al., 1986).

Dantzer and his associates analyzed cellular immune functions in rats of different social status, using the resident-intruder model (Raab et al., 1986). In these experiments they observed a reduced cellular immune reactivity in the subordinate animals (the "loosers"). The stressful situation which develops in the resident-intruder dyads because of the difference in social status of the dominant and subordinate animals apparently modulates immune function. This may imply that the differences in immune reactivity observed in dominant and submissive rats in longterm colonies is not the reflection of intrinsic differences in immune capacity, linked to behavioural characteristics, but rather the result of stress-induced neuroimmunomodulation. This assumption is supported by the serendipitous finding that the leader of a colony, after being deposed, had strongly suppressed immune functions (Koolhaas et al., 1986).

The effect of acute emotional stimuli was investigated by Croiset et al. (1986) using the one-trial passive avoidance test. This paradigm is based on the preference of the rat for the dark and has been developed by Ader et al. (1972). The experimental apparatus consists of a dark compartment, equipped with a grid floor, to which an elevated, illuminated platform is attached. After a number of training trials in which the rat was placed on the platform and allowed to enter the dark compartment, the animal received a single, unavoidable, mild (0.9 mA for 2 sec.) electric footshock immediately upon entering the dark room (learning trial). Learning experience (passive avoidance) as tested 24 hr later is manifested by avoiding re-entering the dark compartment when placed on the platform (retention test). This passive avoidance behaviour is accompanied by psychological stress, presumably related to anxiety. Control animals were subjected to the same experimental procedure but either the electric footshock was not given during the learning trial or the retention test was omitted. The stressful experience produced by the retention test in the experimental animals resulted in diminished T and B cell responses in vitro as compared to the control animals (Fig. 3). Also the antibody response in vivo was less efficient when the rats were immunized immediately after the retention trial.

In the framework of these studies an interesting observation was made. It was noted that the rats in the control group had higher lymphocyte stimulation values and antibody production than the home-cage control animals. Further analysis suggested that the emotional stimulus caused by the experimental "control situation" represents an element of

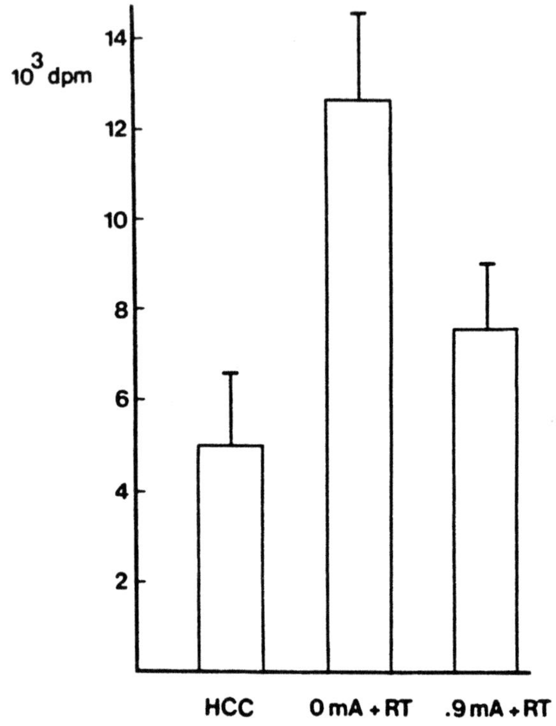

Fig. 3 - Optimal Con A induced proliferation of splenic lymphocytes from home cage control animals (HCC), apparatus control animals (0 mA + RT) and experimental animals (0.9 mA + RT). The results are expressed as dpm ± SEM (n=10-15) (adopted with permission from Croiset et al., 1986)

novelty rather than anxiety. The arousal, associated with exposure to a novel, unexpected situation, apparently results in increased immune reactivity as compared to home-cage control rats. The emotional stress induced by the retention session seems to "neutralize" this effect. These findings are of basic interest since it was found that also in a mild, short-term (5 min) restraint paradigm the lymphocytes of the experimental animals show increased reactivity, both _in vitro_ as well as _in vivo_, as compared to home cage control rats (Berkenbosch et al., 1986). Preliminary data suggest that the immunostimulatory effect of this type of emotional stimulus is mediated via the hypothalamus. Since stress causes a rapid release of pituitary hormones, such as ACTH, prolactin and endorphins, one or several of these hormones may be involved in the observed modulation of immune reactivity.

CONDITIONING OF THE IMMUNE RESPONSE

A very interesting new area of research has been opened up by the pioneer-work of Robert Ader. He showed that in rats (and also in mice) a decrease in immune reactivity can be conditioned by pairing a conditioned stimulus (flavoured drinking water) with an unconditioned stimulus, represented by the injection of the immunosuppressive drug cyclophosphamid. The conditioned animals, when exposed to the conditioning stimulus at a later stage, show a suppressed immune function. This original observation by Ader (Ader and Cohen, 1975) has been confirmed and extended by several groups using different paradigms (reviewed by Ader and Cohen, 1985). An important issue in this area of behavioural immunology is the mechanism(s) underlying conditioned changes in immune responses. Ader and Cohen (1985) are of the opinion that adrenocorticosteroid hormones do not play a central role in mediating the conditioned immunosuppression of antibody responses in rats (see also Ader et al., 1986). However, Gorczynski et al. (1984) have shown that adrenalectomy abolishes the capacity to condition immunosuppression of the antibody response in mice. Interestingly, this author recently reported that the drug cimetidine reverses conditioned immunosuppression at the T cell level (Gorczynski et al., 1985). This strongly suggests that histamine-receptor carrying T suppressor cells are involved in conditioned immune suppression. If this can be confirmed, new avenues can be choosen to influence the immune system linking immunopharmacology with behavioural immunology. Finally, the results obtained by Kelley et al. (1985) should be mentioned. These investigators induced conditioned taste aversion in mice using the non-immunosuppressive drug lithium chloride (as has been done before by Ader et al., see Ader & Cohen, 1985). They observed in conditioned mice a strongly reduced delayed type hypersensitivity immune response. Since the drug used as the unconditioned stimulus has no immunosuppressive effect, and because it was found that the conditioned animals had higher plasma levels of glucocorticoids than the non-conditioned mice, the authors argue that stress rather than conditioning underlies the immunosuppression induced by reexposure of the conditioned animals to the aversive taste.

BRAIN-LYMPHOID AXIS: A TWO-WAY CONNECTION?

In the emerging new discipline Psychoneuroimmunology, the main emphasis has been laid on the influence of the brain and the nervous system on the immune system. From the results of many studies it is obvious that the neuroendocrine system plays an important role in mediating the stress-induced immunomodulation. However, this picture is incomplete. Already several years ago Besedovski et al. (1977) have shown that an on-going immune response increases the firing rate of neurons in the brain. It seems that lymphokines, produced by cells of the immune system can bring about this effect. It furthermore was shown that in the course of an immune response, the corticosteroid level in the blood, and the noradrenaline level in the spleen of the immunized animal increases (Besedovski et al., 1979). These observations imply that not only the brain influences the immune system but that the immune system reciprocally may effect the central nervous system. Recent findings by several groups strongly support this notion. Thus, Blalock et al. (1985) reported that murine lymphocytes, after proper activation with mitogens, produce molecules which on the basis of reactivity towards (monoclonal) antibodies, apparently are very similar if not identical to pituitary hormones of the neuroendocrine circuit. Indeed, in a limited number of studies the molecular basis for this finding has been obtained (Blalock, 1986; Westly et al., 1986). It therefore seems reasonable to expect that lymphocyte derived "neurotransmittors" can influence behaviour. The clinical relevance of this concept has yet to be established.

ACKNOWLEDGEMENT

This paper includes data from research carried out in the authors' laboratories. Part of these investigations have been supported by grants from the Dutch Asthma Foundation (grant no. 84.21) and from Organon International, Oss, The Netherlands.

REFERENCES

Ader, R. (Ed.) 1981. Psychoneuroimmunology. (Academic Press, New York)

Ader, R., Weijnen, J.A.W.M. and Moleman, P. 1972. Retention of a passive avoidance response as a function of the intensity and duration of electric shock. Psychonomic Science, 26, 125-128.

Ader, R. and Cohen, N. 1975. Behaviorally conditioned immunosuppression. Psychosom. Med., 37, 333-340.

Ader, R. and Cohen, N. 1985. CNS-immune system interactions: conditioning phenomena. Behav. Brain Sci., 8, 379-426.

Ader, R., Cohen, N.A. and Grota, L.J. 1986. Adrenocortical steroids in the conditioned suppression and enhancement of immune responses. In "Hormones and Immunity" (Ed. I. Berczi) (Elsevier, New York), in press.

Ballieux, R.E. (Ed.). 1984a. Psychoneuroimmunology and breakdown in adaptation: interactions within the central nervous system, the immune and endocrine systems. In "Breakdown in human adaptation to stress", vol. II (Martinus Nijhoff Publ., The Hague) pp. 623-739.

Ballieux, R.E. 1984b. Stress and immune response: parameters and markers. In "Breakdown in human adaptation to stress", vol. II (Martinus Nijhoff Publ., The Hague) pp. 732-739.

Ballieux, R.E., Croiset, G., Veldhuis, H.D., de Wied, D., Berkenbosch, F., Tilders, F.J.H., Smelik, P., Koolhaas, J., Bohus, B. and Heijnen, C.J. 1986. The influence of various forms of stress on the immune system. Poster, presented at the 6th International Congress of Immunology, Toronto, Canada, 6-11 July 1986.Abstract no. 3.63.38.

Berkenbosch, F., Heijnen, C.J., Croiset, G., Bevers, C., Ballieux, R.E., Binnekade, R. and Tilders, F.J.H. 1986. Endocrine and immunological changes in response to acute stress. In "The enkephalins and endorphins: stress and the immune system" (Ed. N.P. Plotnikoff, R.E. Faith, A.J. Murgo and R.A. Good). (Plenum Press, New York).

Besedovski, H.O., Sorkin, E., Felix, D. and Haas, H. 1977. Hypothalamic changes during the immune response. Eur. J. Immunol., 7, 323-325.

Besedovski, H.O., del Rey, A., Sorkin, E., Da Prada, M. and Keller, H.H. 1979. Immunoregulation mediated by the sympathetic nervous system. Cell. Immunol. 48, 346-355.

Blalock, J.E., Harbour-McMenamin, D. and Smith, E.M. 1985. Peptide hormones shared by the neuroendocrine and immunological systems. J. Immunol., 135, 858s-861s.

Blalock, J.E. 1986. Production and action of lymphocyte-derived neuroendocrine peptide hormones - summary. Progress in Immunology, 6, in press.

Croiset, G., Veldhuis, H.D., Ballieux, R.E., de Wied, D. and Heijnen, C.J. 1986. The impact of mild emotional stress induced by the passive avoidance procedure on immune reactivity. Ann. N.Y. Acad. Sci., in press.

Goetzl, E.J. (Ed.) 1985. Neuromodulation of immunity and hypersensitivity. J. Immunol., 135, 739s-863s.

Gorczynski, R., MacRae, S. and Kennedy, M. 1984. Factors involved in the classical conditioning of antibody responses in mice. In "Breakdown in human adaptation to stress", vol. II (Martinus Nijhoff Publ., The Hague) pp. 704-713.

Gorczynski, R., Kennedy, M. and Ciampi, A. 1985. Cimetidine reverses tumor growth enhancement of plasmacytoma tumors in mice demonstrating conditioned immunosuppression. J. Immunol., 134, 4261-4266.

Guillemin, R., Cohn, M. and Melnechuk, T. (Eds) 1985. Neural modulation of immunity. (Raven Press, New York).

Heijnen, C.J., Bevers, C., Kavelaars, A. and Ballieux, R.E. 1986a. Effect of alpha-endorphin on the antigen-induced primary antibody response of human blood B cells in vitro. J. Immunol., 136, 213-216.

Heijnen, C.J., Croiset, G., Zijlstra, J. and Ballieux, R.E. 1986b. Modulation of lymphocyte function by endorphins. Ann. N.Y. Acad. Sci., in press.

Johnson, H.M., Smith, E.M., Torres, B.A. and Blalock, J.E. 1982. Regulation of the in vitro antibody response by neuroendocrine hormones. Proc. Natl. Acad. Sci. (USA), 79, 4171-4174.

Kelley, K.W., Dantzer, R., Mormede, P., Salmon, H. and Aynaud, J.M. 1985. Conditioned taste aversion suppresses induction of delayed-type hypersensitivity immune reactions. Physiol. Behav., 34, 189-193.

Koolhaas, J., Kamstra, A., Bohus, B. and Heijnen, C.J. 1986. Social structure and state of the immune system in male rats. This volume.

Monjan, A.A. 1981. Stress and immunologic competence: studies in animals. In "Psychoneuroimmunology" (Ed. R. Ader). (Academic Press, New York) pp. 185-228.

Oliverio, A. 1986. This volume.

Plotnikoff, N.P., Faith, R.E., Murgo, A.J. and Good, R.A. (Eds) 1986. Enkephalins and endorphins: stress and the immune system. (Plenum Publ. Corp., New York).

Raab, A., Dantzer, R., Michaud, B., Mormede, P., Taghzouti, K., Simon, H. and Le Moal, M. 1985. Behavioural, physiological and immunological consequences of social status and aggression in chronically coexisting resident-intruder dyads of male rats. Physiol. Behav., 36, 223-228.

Westly, H.J., Kleiss, A.J., Kelley, K.W., Wong, P.K.Y. and Yuen, P.-H. 1986. Newcastle disease virus-infected splenocytes express the pro-opiomelanocortin gene. J. Exp. Med., 163, 1589-1594.

EFFECTS OF BEHAVIOURAL AND PHYSICAL STRESSORS ON IMMUNE RESPONSES

H. S. Siegel
Department of Poultry Science
The Pennsylvania State University
University Park, PA 16802

ABSTRACT

Research in animals over the past several decades had indicated that the relationship between the effects of physical and behavioral stressors and the immune response was unidirectional and thus, imposition of unfavorable environments could suppress normal antibody and cell mediated responses. Evidence indicated that a significant part of this suppression was mediated through hypothalamic-pituitary-adrenal neuroendocrine pathways. However, research now indicates that this pathway may not be unidirectional; that information also flows from elements of the immune system toward central control. Extrapituitary, extrahypothalamic "ACTH-like" and "β-endorphin-like" peptides are produced by cells of the immune system that may be initiators of immunity and interact with central mechanisms.

INTRODUCTION

For the past several decades it has been assumed that the relationship between the effects of unfavorable environments (popularly defined as "stress") and the immune system of the animal is unidirectional. That is, if one imposes a physical or behavioral stressor of sufficient intensity or length on an animal, an endocrine response, usually defined in terms of the hypothalamic-pituitary-adrenal pathway, will suppress the ability of the animal to initiate or maintain optimum immunological responses. There seems to be considerable evidence to support this concept.

The direct effects of corticosteroids, or the indirect effects of corticotropin (ACTH) or stress, on lymphoid tissues are well known. These effects include reductions in lymphatic tissue mass (i.e., thymus, spleen, bursa of Fabricius) (Garren and Shaffner, 1954, 1956; Glick, 1957, 1967; Siegel and Beane, 1961), a depression in the number of circulating lymphocytes and an increase in the neutrophilic or heterophilic granulocytes (Dougherty and White, 1944; Gross and Siegel, 1983).

The mechanism for the effects of corticosteroids on the immune system have become clear. *In vitro* studies show that corticosteroids bind to specific cytoplasmic receptors in lymphatic cells, and that the steroid-receptor complex passes into the nucleus of the cell to alter enzymatic activity and influence nucleic acid metabolism (Thompson and Lippman, 1974; Sullivan and Wira, 1979). *In vivo* studies showed that ACTH or exposure to high temperature increased the amount of endogenously produced corticosteroid bound in the lymphocytes (Fig. 1A, 1B). Evidence suggests that the immune suppressive effects of corticosteroids are on thymus-derived lymphocytes (T-cells) (Meyer et al., 1964; Sato and Glick, 1970; Pardue and Thaxton, 1984).

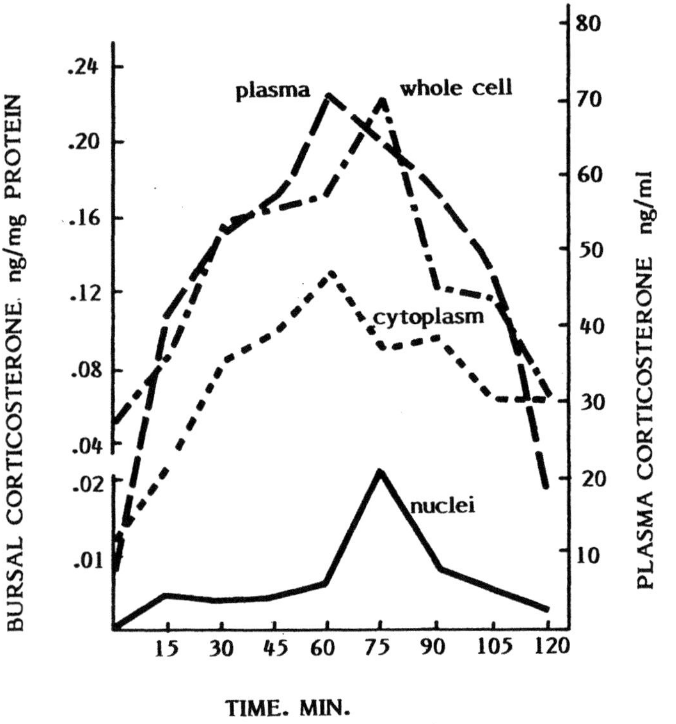

Fig. 1A. Effect of ACTH on the binding of endogenously produced corticosteroid in the lymphocytes of the bursa of Fabricius of 7-week-old White Rock chickens (Data of Gould and Siegel, 1981).

Physical Stressors

Studies with several domestic species show that a variety of stressors are capable of reducing levels of circulating antibody and suppressing cell mediated immunity (CMI). For example, when growing chickens were immunized against a variety of antigens (Ag), such as Salmonella pullorum (SP), sheep red blood cells (SRBC) or bovine serum albumin (BSA) and subjected to high environmental temperature (four 42°C episodes of 0.5 hr over a 4-hr

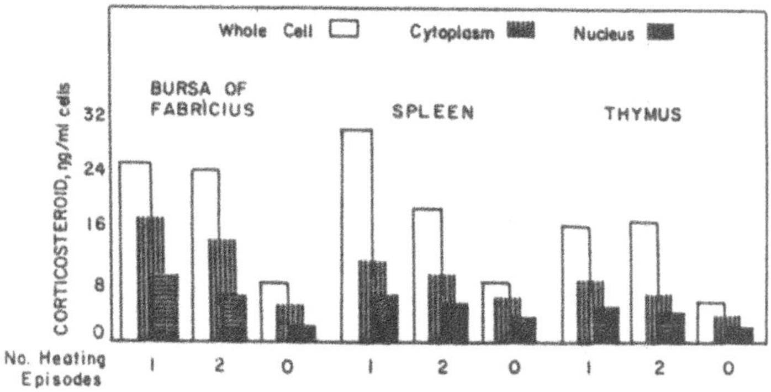

Fig. 1B. Effect of heating episodes on the binding of endogenously produced corticosteroid into the lymphocytes of three organs of 7-week-old White Rock chickens (Data of Siegel and Gould, 1982).

period), humoral antibodies to these antigens were depressed within 12 hr after the exposure (Thaxton and Siegel, 1970). The reduction of plasma immunoglobulin G in calves chronically exposed to high temperature suggests a similar response may be found in bovine species (Kelley et al., 1982), although a longer, more continuous heat exposure may be required (Kelley et al., 1981) than in fowl. Exposure to high temperature has also been found to reduce passive (maternally transferred) immunity in calves (Stott et al., 1976) and CMI in fowl (Regnier and Kelley, 1981).

Hypothermic stress may also effectively reduce immune responses. For example, cold exposure lowered colostral antibodies in calves (Blecha and Kelley, 1981) and reduced lung clearance rates of E. Coli in neonatal pigs (Curtis et al., 1976). In avian species it was found that low temperature exposure reduced immunity to Mycoplasma meleagrida in turkeys (Brown and Nestor, 1973) and CMI responses in chickens (Blecha and Kelley, 1981).

The immunodepressive effects were shown to be mediated, in part, by corticosteroids, because pre-stress treatment with metyrapone, a suppressor of corticosterone syntheses, modifies the stress effect (Thaxton and Siegel, 1973). Adrenal involvement in the suppression of CMI by immobilization stress has also been shown (Blecha et al., 1982).

However, experimental results showing that physical stressors suppress immunological responses have not always been consistent. Stress-induced immunosuppression, although often observed in CMI reactions is frequently not observed in antibody reactions (Thaxton and Briggs, 1972; Morgan et al., 1976; Subba Rao and Glick, 1977; Regnier et al., 1980). The inability of corticosteroids to alter certain CMI responses suggests that not all T-cell populations are influenced in the same manner (Blecha et al., 1982), or other substances, also involved in stress responses such as catecholamines (Edens and Siegel, 1975) may also suppress the corticotropin-releasing ability of the hypothalamus (Jones and Hillhouse, 1977).

Variations in the degree that stress alters immune responses depend on several factors:

1. Genetics - Genetic differences in the ability of animals to mount an antibody response have been shown (Biozzi et al., 1979; Van der Zijpp, 1983). There are also differences among fowl in the degree of corticosteroid responsiveness to ACTH (Edens and Siegel, 1975; Gross and Siegel, 1973). Because the effects of stress may depend on the degree that corticosteroids bind in the cell, it is of interest that lymphocytes of birds selected for high response to ACTH bind greater amounts of endogenously produced corticosteroid after ACTH activation than low stress responders (table 1).

2. Nutritional - Moderate (80% of *ad lib*) restrictions or intake of a balanced diet had little effect on immunological status of birds acclimatized to 25°C; a similar feed restriction stimulated increased hemagglutinin (HA) responses in birds acclimatized to 15-35°C or fluctuating 10-20°C or 30-40°C (Henken et al., 1983). However, restrictions of 80% of *ad lib* at temperatures below 10°C, or subjection to sudden 10°C changes in temperature depressed HA titers (Henken et al., 1982). Severe energy or amino acid restriction (1/3 of metabolizable energy or 1/3 lysine-methionine requirements) depressed HA responses to SRBC (Glick et al., 1981) or delayed hypersensivity reactions to human gamma globulin (Glick et al., 1983) in chickens; however, splenic lymphocytes appeared capable of normal graft vs. host responses in the nutrient-deficient birds. Adequate diets returned normal immune status within 2 weeks.

TABLE 1 Effect of ACTH on corticosteroid binding in thymus cells of Athens Randombred chickens selected for high and low agglutinin response to ACTH.

	ACTH Agglutinin Responses		
	High		Low
Whole cell, ng/ml cells			
ACTH	37.3 ± 9.2[a]		19.7 ± 3.2[b]
Saline	13.3 ± 0.7	***	13.6 ± 0.3
Cytoplasm, ng/ml cells			
ACTH	16.2 ± 2.6[a]		7.7 ± 1.2[b]
Saline	1.6 ± 0.3	***	1.9 ± 0.3
Nucleus, ng/ml cells			
ACTH	16.0 ± 3.9[a]		9.5 ± 1.4[b]
Saline	1.3 ± 0.2	***	1.6 ± 0.2

[a,b]Means ± SE within rows with different superscripts are significantly different ($P<.05$).

[***]Differences between ACTH and saline-treated groups are significant ($P<0.0001$).

Specific vitamins and trace mineral deficiences have also been found to suppress immune responses in animals. Vitamine E deficiency depresses normal humoral responses to SRBC in chickens (Marsh et al., 1971) and low plasma levels of vitamin C have been associated with reduced immune function in humans (Pauling, 1976), mice (Prinz et al., 1980) and fowl (Pardue and Thaxton, 1984). Corticosteroids cause a depletion of plasma ascorbic acid and dietary supplementation corrects this. Selenium deficiency alters the immune status of cattle (Arthur et al., 1982) and fowl (Colgano et al., 1984). These factors appear to provide protective antioxidants during macrophage-lymphocyte interaction that are oxidative in nature (Brune and Spitznagel, 1973; Bellavite et al., 1977; Weiss et al., 1983).

3. Antigen concentration - Antigen concentrations also may influence the effect that stressors have on antibody responses. For example, ACTH or acute high temperature suppressed agglutinating antibody against *S. pullorum* when certain critical levels of antigen were used, but failed to

suppress these responses at higher antigen concentrations (Siegel et al., 1983, 1984) (Fig. 2A, 2B)

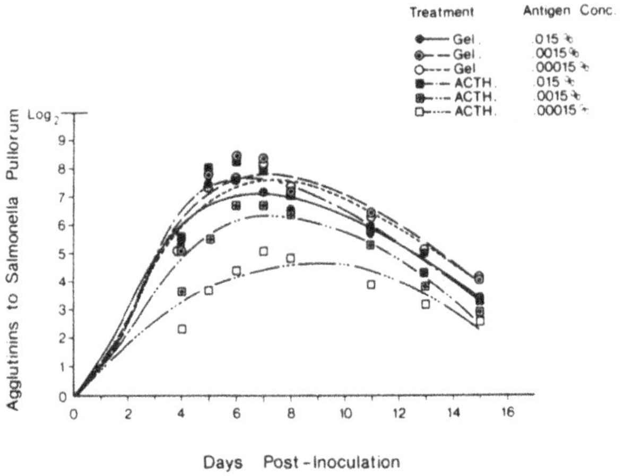

Fig. 2A. Polynomial plots showing the effect of antigen concentration on ACTH suppression of agglutinin titers to S. pullorum (Siegel et al., 1983). Gel. = gelatine, 16% in H_2O.

However, although the immunosuppressant effects of vitamin E were more readily observed at low antigen doses (Marsh et al., 1981), the effects of energy or amino acid deficiency appeared to be more demonstrable at high antigen dosage (Glick et al., 1981). Gross (1979) found genetic differences in immune response to behavioral stress were more readily observed at low antigen levels, but van der Zijpp (1983) could not demonstrate this result with commercial stock. The results suggest that an optimal antigen concentration may allow more genetic loci to be operational (Biozzi et al., 1979).

4. Interactions of non-specific and specific responses - Physiological responses to the environment can be divided into specific responses - those in which a particular condition elicits a specific response - and non-specific responses - where, regardless of the offending stimulus, the organism responds in a generalized manner - a state of general stress.

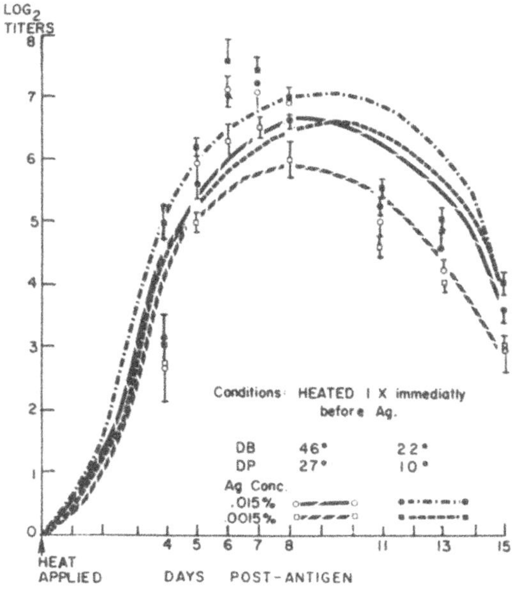

DB = Dry Bulb temp,
DP = Dewpoint temp.

Fig. 2B. Interaction of high temperature exposure and antigen concentration on serum agglutinin responses to S. pullorum in White Rock males (Siegel and Latimer, 1984).

For example, homeotherms exposed to temperature below lower critical temperatures, increase heat production, i.e., increase metabolic rate, to maintain body temperature. Primary regulators are thyroid stimulating hormones that activate the thyroid by increasing adenyl cyclase activity and thus increase cyclic-AMP activity (Dumant et al., 1971). In addition to the general increase in metabolism that stimulates protein synthesis in general, cyclic-AMP specifically increases synthesis of antibody protein (Braun et al., 1971). Therefore, increases in antibody often accompany exposure to moderately low temperature in chickens (Subba Rao and Glick, 1977), pigs (Blecha and Kelley, 1981) and calves (Kelley et al., 1981, 1982).

However, if heat loss to the environment exceeds heat production, then metabolic processes decline and stress responses assume importance. Such conditions can occur when: (1) ambient temperatures fall below the animal's ability to produce sufficient heat; (2) normal insulative abilities are broached (Nir et al., 1975; Siegel, 1980); (3) homeothermic capacity is immature, as in neonates (Blecha and Kelley, 1981); (4) insufficient energy intake or reserves for required level of heat production (Henken et al., 1982, 1983).

Behavioral Stressors

Social or behavioral environments are also activators of stress responses in animals, and like physical stressors they are capable of reducing immune responses. Domestic fowl in stressful social environments have been shown to have reduced antibodies against a variety of antigens such as SRBC, S. pullorum, E. Coli, Newcastle disease virus, and Marek's disease virus (Gross and Siegel, 1973, 1975; Siegel and Latimer, 1975; Thompson et al., 1980; Edens et al., 1983) (see Fig. 3).

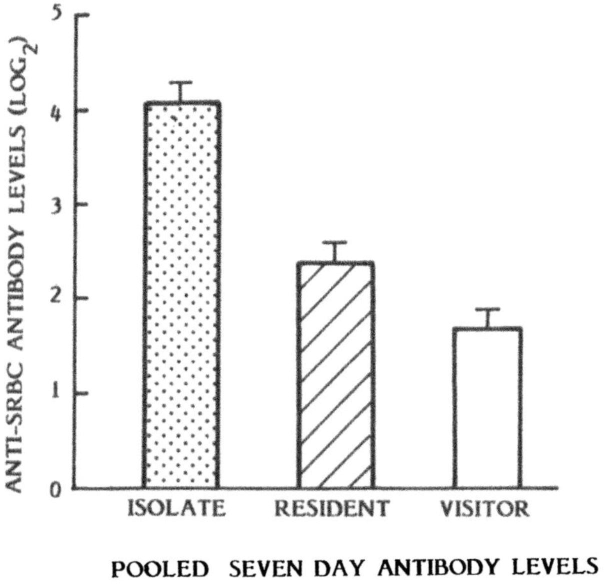

Fig. 3. Effect of stressful social environment on hemagglutinin response to chukar red blood cells in Japanese quail (Edens et al., 1983). Isolate: birds were handled daily, but no new members were introduced. Residents: groups of birds into which a new member was moved daily. Visitors: the new member that was introduced; visitors were moved daily from group to group for 11 days.

Similarly, with mammals, increased social strife reduces immune responses in rats (Soloman, 1969). Sows prevented from performing normal maternal behavior have lower levels of agglutinating antibodies against foreign cells (Oosterlee et al., 1980).

As noted with physical stressors, immunological responses to behavioral stressors have heritable components. When introduced as strangers into established flocks, cockerels from less aggressive families have more rapid regression of agglutinin titers against S. pullorum than those from more aggressive families (Siegel and Latimer, 1975). Genetic lines of chickens have been selected for high (HPC) or low (LPC) plasma corticosteroid response to social stress (Gross and Siegel, 1973). Not only did HPC line produce lower antibody levels to E. Coli and SRBC, but effects of social stress were additive (Gross and Siegel, 1975). A parallel to this genetic environment interaction has appeared in lines responsive to ACTH (Edens and Siegel, 1975). Birds with pea-combs are almost subordinate to those with single combs. Approximately one-third of the high ACTH-responding line are pea-combed individuals and these had high plasma corticosteroid levels and low antibody responses to bacterial antigens (Gross and Siegel, 1975).

Thus we have seen that physical and behavioral stressors are capable of suppressing antibody and cell mediated immunity in animal species, and that important mediators of this induced immunosuppression are adrenal corticosteroids. Moreover, immunosuppressive and immunostimulatory effects of neurohumoral peptides, such as ACTH, α-melanocyte stimulating hormones (α-MSH) or β-endorphin, have been reported (Fabris et al., 1971; McCain et al., 1982; Krieger, 1983).

Evidence for non-unidirectional relationships

Evidence is now accumulating that the relationships between environmental stimuli and immune responses are not unidirectional; that information flows from elements of the immune system back through neurohumoral pathways toward the central nervous system. Besedovsky et al. (1975), for example, have shown that an increase in plasma corticosterone follows one day after immunization of mice with equine erythrocytes, or four days after immunization of rats with SRBC. These responses were concomitant with a decline in blood thyroxine and were followed by an increase in splenic plaque-forming ability. It was subsequently shown that immunization of rats with SRBC also brought about significant declines in splenic norepinephrine (Besedovsky et al., 1979). In fowl, there is an increase in metabolic heat production about four days after immunization with SRBC that coincides with the appearance of 2-mercaptoethanol resistant (IgG) antibody (Siegel et al., 1982). These responses, which are proposed as being related to lymphokine production by antigen-stimulated lymphocytes (Besedovsky et al., 1981), appear to be part of a feed-back communication between the immune system and neurohumoral centers.

However, extrapituitary and extrahypothalamic "ACTH-like" peptides, derived from the larger precursor, pro-opiomelanocortin (POMC) have also been described (Krieger, 1983). Moreover, interferon preparations from Newcastle disease virus (NDV)-infected human leukocytes contain biologic and immunologic β-endorphin and ACTH-like activities (Smith and Blalock, 1981). The ACTH-like activity was expressed following pepsin or acid treatment that produced two smaller peptides from the larger molecule. The injection of hypophysectomized mice with NDV caused a rise in serum corticosterone 8 hours after the injection, and the splenic leukocytes of the injected mice displayed positive immunofluorescence to an antibody developed against the 1-13 ACTH-amide (Smith et al., 1982). These results strongly suggested that an ACTH-like compound was being produced by the antigenetically stimulated lymphocyte, which was acting directly on the adrenal gland without mediation by the pituitary.

The results with mice, however, do not exclude the possibility that NDV-stimulation might also be from hypothalamic origin. In a series of experiments with chickens, we were able to establish direct leukocyte-adrenal activity. This result was accomplished by co-incubating splenic leukocytes taken from chickens 1 hour after they were given injections of heat-killed S. pullorum antigen (Siegel et al., 1985).

Figure 5A shows that serum corticosteroid levels begin to rise within 15 minutes, reach a maximum in 2 hours, and fall to control levels by 12 hours after antigen injection. Figure 5B shows that serum taken from cockerels 1 hour after antigen injection contains high levels of ACTH-like activity unless it is removed by stripping with activated charcoal and precipitated silica. Figure 5C shows that splenic leukocytes taken from chickens 1 hour after antigen injection, and co-incubated with isolated adrenal cells from non-antigen treated birds, display significantly higher adrenal corticosteroid stimulating activity.

These results suggest that the secretion of ACTH-like, and potentially β-endorphin-like, compounds by antigen-stimulated lymphocytes may be important in the process of initiation of immunity. How environmental stimuli may interact with these processes during the early stages of the response remains to be elucidated.

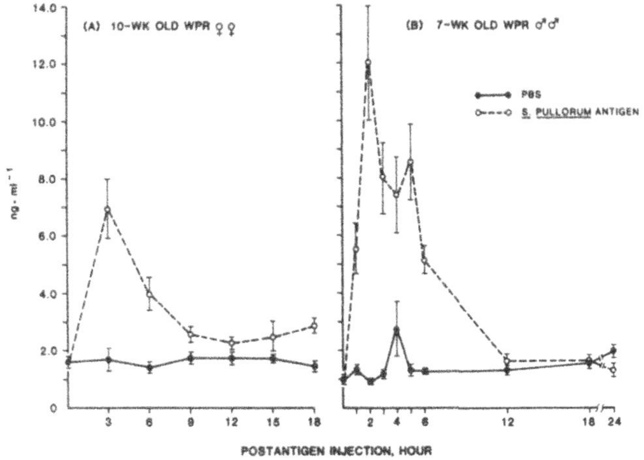

Fig. 5A. Effect of injection of heat-killed S. pullorum antigen and phosphate buffered saline (PBS) on serum corticosteroid levels (ordinate) in White Rock chickens (Siegel et al., 1985).

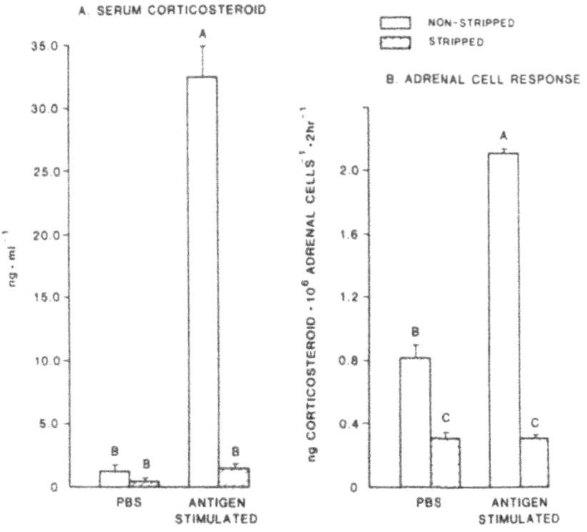

Fig. 5B. Serum corticosteroid and adrenal cell responses to sera from 4-week-old White Leghorn cockerels injected with S. pullorum antigen, and effect of stripping with activated charcoal and Quso-G32 (Siegel et al., 1985).

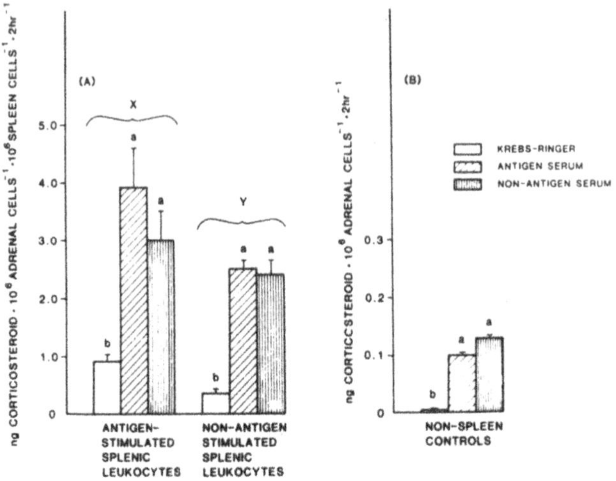

Fig. 5C. Effect of 1 hr in vivo stimulation with S. pullorum antigen on the corticosteroid production of isolated adrenal cells by splenic leukocytes (Siegel et al., 1985).

REFERENCES

Arthur, J. R., Boyne, R., Okolow-Zubkowska, M. J. and Hill, H. A. O. 1982. Neutrophils from Se and Cu deficient cattle. In, Trace Elements in Metabolism in Man and Animals. Ed. J. M. Cawthorne, J. Mc C. Howell, and C. L. White.

Bellavite, P., Dri, P., Bisiacchi, B. and Patriarca, P. 1977. Catalase deficiency in myeloperoxidase deficience polymorphonuclear leucocytes from chickens. FEBS Letters, 81, 73.

Besedovsky, H., Sorkin, E., Keller, M. and Muller, J. 1975. Changes in blood hormone levels during the immune response. Proc. Soc. Exp. Biol. Med., 150, 466-470.

Besedovsky, H. P., Del Rey, A., Sorkin, E., DaPrada, M. and Keller, H. H. 1979. Immunoregulation mediated by the sympathetic nervous system. Cell Immunol., 48, 346-355.

Besedovsky, H. O., Del Rey, A. and Sorkin, E. 1981. Lymphokine-containing supernatants from con A-stimulated cells increase corticosterone blood levels. J. Immunol., 126, 385-387.

Biozzi, G., Mounton, D., Santa-Anna, O. A., Passos, H. C., Gennari, M., Reis, M. H., Ferreira, V. C. A., Heumann, A. M., Bouthaillier, Y., Ibanez, O. M., Stiffel, C. and Siqureia, M. 1979. Genetics of immunoresponsiveness to natural antigens in the mouse. Curr. Top. Microbiol. Immunol., 5, 31-98.

Blecha, F. and Kelley, K. W. 1981. Cold stress reduces the acquisition of colostral immunoglobulin in piglets. J. Anim. Sci., 52, 594-600.

Blecha, F., Kelley, K. W. and Satterlee, D. G. 1982. Adrenal involvement in the expression of delayed-type hypersensitivity in SRBC and contact sensitivity to DNFB in stressed mice. Proc. Soc. Exp. Biol. Med., 169, 247-252.

Braun, W., Masaaki, I., Winchurch, R. and Webb, D. 1971. On the role of cyclic AMP in immune responses. Ann. N.Y. Acad. Sci., 185, 417-422.

Brown, K. I. and Nestor, K. E. 1973. Some physiological responses of turkeys selected for high and low adrenal response to cold stress. Poultry Sci., 52, 1948-1954.

Brune, K. and Spitznagel, J. K. 1973. Peroxidaseless chicken leucocytes: isolation and characterization of antibacterial granules. J. Infect. Dis., 127, 84-88.

Colgano, G. L., Jensen, L. S. and Long, P. L. 1984. Effect of selenium on peripheral blood leucocytes of chickens infected with Emeria. Poultry Sci., 63, 1136-1144.

Curtis, S. E., Kingdon, D. A. and Drummond, J. G. 1976. Effects of age and cold on pulmonary bacterial clearances in the young pig. Am. J. Vet. Med., 37, 299-301.

Doughtery, T. F. and White, A. 1944. Influence of hormones on lymphoid tissue structure and function. The role of pituitary adrenotrophic hormone in the regulation of lymphocytes and other cellular elements of blood. Endocrinology, 35, 1-12.

Dumont, J. E., Willems, C., Van Sande, J. and Neve, P. 1971. Regulation of the release of thyroid hormones: the role of cyclic AMP. Ann. N.Y. Acad. Sci., 185, 291-316.

Edens, F. W. and Siegel, H. S. 1975. Adrenal responses in high and low ACTH response lines of chicks during acute heat stress. Gen. Comp. Endocrinol., 25, 64-73.

Edens, F. W., Thaxton, P., Morgan, G. W. and Parkhurst, C. R. 1983. Grouping in Japanese quail. 2. Suppression of the humoral immunity. Poultry Sci., 62, 2479-2485.

Fabris, N. W., Pierpaoli, W. and Sorkin, E. 1971. Hormones and immunological capacity. IV. Restorative effects of developmental hormones or of lymphocytes on the immunodeficiency syndrome in dwarf mice. Clin. Exp. Immunol., 9, 227-240.

Garren, H. W. and Shaffner, C. S. 1954. Factors concerned in the response of young New Hampshires to muscular fatigue. Poultry Sci., 33, 1095-1104.

Garren, H. W. and Shaffner, C. S. 1956. How the period of exposure to different stress stimuli affects the endocrine and lymphatic gland weights of young chicks. Poultry Sci., 35, 266-272.

Glick, B. 1957. Experimental modification of the growth of the bursa of Fabricius. Poultry Sci., 36, 18-24.

Glick, B. 1967. Antibody and gland studies in cortisone and ACTH-injected birds. J. Immunol., 98, 1076-1084.

Glick, B., Day, E. J. and Thompson, D. 1981. Calorie-protein deficiencies and the immune response of the chicken. I. Humoral immunity. Poultry Sci., 60, 2494-2500.

Glick, B., Taylor, R. L., Jr., Martin, D. E., Watabe, D., Day, E. J. and Thompson, D. 1983. Calorie-protein deficiencies and the immune response of the chicken. II. Cell-mediated immunity. Poultry Sci., 62, 1889-1893.

Gould, N. R. and Siegel, H. S. 1981. Viability of and corticosteroid binding in lymphoid cells of various tissues after corticotropin injection. Poultry Sci., 60, 891-893.

Gross, W. B. 1979. Comparison of dose effect of sheep red blood cells on antibody response in two lines of chickens. Avian Dis., 23, 526-530.

Gross, W. B. and Siegel, H. S. 1983. Evaluation of heterophili:lymphocyte ratio as a measure of stress in chickens. Avian Dis., 27, 972-979.

Gross, W. B. and Siegel, P. Bl 1973. Effects of social stress and steroids on antibody production. Avian Dis., 17, 807-815.

Gross, W. B. and Siegel, P. B. 1975. Immune response to Escherichia coli. Am. J. Vet. Res., 36, 568-571.

Henken, A. M., Groote Schaarsberg, A. M. J. and Nieuwland, M. G. B. 1983. The effect of environmental temperature on immune response and metabolism of the young chickens. 3. Effect of environmental temperature on the humoral immune response following injection of sheep red blood cells. Poultry Sci., 62, 51-58.

Henken, A. M., Verstegen, M. W. A., van der Hel, W. and Knol, E. F. 1982. The effect of environemtnal temperature on immune response and metabolism of the young chickens. 6. Effect of environmental temperature on heat production in pullets in relation to feeding level. Proc. 9th Symp. Energy Metab. Farm. Anim., Lillihammer, Norway.

Jones, M. T. and Hillhouse, E. W. 1977. Neurotransmitter regulation of corticotropin-releasing factor, in vitro. Ann. N.Y. Acad. Sci., 297, 536-549.

Kelley, K. W., Osborne, C. A., Evermann, J. F., Parish, S. M. and Hinrichs, D. S. 1981. Whole blood leukocyte vs. separated mononuclear cell blastogenesis in calves: Time-dependent changes after shipping. Can. J. Comp. Med., 45, 248-258.

Kelley, K. W., Osborne, C. A., Evermann, J. F., Parish, S. M. and Gaskins, C. T. 1982. Effect of chronic heat and cold stressors on plasma immunoglobulin and mitogen-induced blastogenesis in calves. J. Dairy Sci., 65, 1514-1528.

Krieger, D. T. 1983. Brain peptides: What, when, and why? Science (Washington, D.C.), 222, 975-985.

McCain, H. W., Lamster, I. B., Bonnone, J. M. and Grbic, J. T. 1982. B-endorphin modulates human immune activity via non-opiate receptor mechanisms. Life Sci., 31, 1619-1624.

Marsh, J. A., Dietert, R. R. and Combs, G. F., Jr. 1981. Influence of dietary selenium and vitamin E on the humoral immune response of the chick. Proc. Soc. Exp. Biol. Med., 166, 228-231.

Meyer, R. K., Aspinall, R. L., Graetzer, M. A. and Wolfe, H. R. 1964. Effect of corticosterone on the skin homograft reaction and on the precipitin and hemagglutinin production in thymectomized chickens. J. Immunol, 92, 466-450.

Morgan, G. W., Thaxton, P. and Edens, F. W. 1976. Reduced symptoms of anaphlaxis in chickens by ACTH and heat. Poultry Sci., 55, 1498-1504.

Nir, I., Yam, D. and Perek, M. 1975. Effects of stress on the corticosterone content of the blood plasma and adrenal gland of intact and bursectomized Gallus domesticus. Poultry Sci., 54. 2101-2114.

Oosterlee, C. C., Van Dijk, J. H. C., Metz, J. H. M. and de Wilt, J. C. 1980. Influence of housing systems on the immune response of sows, colostral transfer of antibodies and behavior of piglets. Proc. EAAP Conf., Munich, FRG.

Pardue, S. L. and Thaxton, J. P. 1984. Evidence for amelioration of steroid-mediated immunosuppression by ascorbic acid. Poultry Sci., 63, 1262-1269.

Pauling, L. 1976. Vitamin C, The Common Cold and The Flu. W. H. Freeman and Co., San Francisco, CA.

Prinz, W., Block, J., Gilich, G. and Mitchell, G. 1980. A systematic study of the effect of vitamin C supplementation on humoral immune response in ascorbate-dependent mammals. Int. J. Vit. Nutr. Res., 50, 294-301.

Regnier, J. A. and Kelley, K. W. 1981. Heat- and cold-stress suppresses in vivo and in vitro cellular immune responses of chickens. Am. J. Vet. Res., 42, 294-299.

Regnier, J. A., Kelley, K. W. and Gaskins, C. T. 1980. Acute thermal stressors and synthesis of antibodies in chickens. Poultry Sci., 59, 985-990.

Sato, K. and Glick, B. 1970. Antibody and cell mediated immunity in corticosteroid-treated chicks. Poultry Sci., 49, 982-991.

Siegel, H. S. 1980. Physiological stress in birds. Bioscience, 30, 529-534.

Siegel, H. S. and Beane, W. L. 1961. Time responses to single intramuscular doses of ACTH. Poultry Sci., 40, 216-219.

Siegel, H. S. and Gould, H. R. 1982. High temperature and corticosteroid in the lymphocytes of domestic fowl (Gallus domesticus). Gen. Comp. Endocrinol., 48, 348-354.

Siegel, H. S., Gould, N. R. and Latimer, J. W. 1985. Splenic leukocytes from chickens injected with Salmonella pullorum antigen stimulate production of corticosteroids by isolated adrenal cells. Proc. Soc. Exp. Biol. Med., 178, 523-530.

Siegel, H. S. and Latimer, J. W. 1975. Social interactions and antibody titers in young male chickens. Anim. Behav., 23, 323-330.

Siegel, H.S. and Latimer, J.W., 1984. Interaction of high temperature and Salmonella pullorum antigen concentration on serum agglutinin and corticosteroid responses in White Rock chickens. Poultry Sci, 63, 2483-2491.

Siegel, H. S., Latimer, J. W., and Gould, N. R. 1983. Concentration of Salmonella pullorum antigen and the immunosuppressive effect of ACTH in growing chickens. Poultry Sci., 62, 897-903.

Smith, E. M. and Blalock, J. E. 1981. Human lymphocyte production of corticotropin and endorphin-like substances: Association with leukocyte interferon. Proc. Natl. Acad. Sci. USA, 78, 7530-7534.

Smith, E. M., Meyer, W. J. and Blalock, J. E. 1982. Virus-induced corticosterone in hypophysectomized mice: A possible lymphoid-adrenal axis. Science (Washington, D.C.), 218, 1311-1312.

Soloman, G. F. 1969. Stress and antibody response in rats. Intl. Arch. Alelrgy, 35, 97.

Stott, G. H., Wiersma, F., Menefee, B. E. and Radwanski, F. R. 1976. Influence of environmenton passive immunity in calves. J. Dairy Sci., 59. 1306-1312.

Subba Rao, D. S. V. and Glick, B. 1977. Effect of cold exposure on the immune response of chickens. Poultry Sci., 56, 992-996.

Sullivan, D. A. and Wira, C. R. 1979. Sex hormone and glucocorticoid receptor in the bursa of Fabricius of immature chickens. J. Immunol., 122, 2617-2623.

Thaxton, P. and Briggs, D. M. 1972. Effect of immobilization and formaldehyde on immunological responsiveness in young chickens. Poultry Sci., 51, 342-344.

Thaxton, P. and Siegel, H. S. 1970. Immunodepression in young chickens by high environmental temperature. Poultry Sci., 49, 202-205.

Thaxton, P. and Siegel, H. S. 1973. Modification of high temperature and ACTH mediated immunosuppression by metryrapone. Poultry Sci., 52, 618-624.

Thompson, D. L., Elgert, K. D., Gross, W. B. and Siegel, P. B. 1980. Cell-mediated immunity in Marek's disease virus=infected chickens genetically selected for high and low concentrations of plasma corticosterone. Am. J. Vet. Res., 41, 91-96.

Thompson, E. B. and Lippman, M. E. 1974. Mechanism of action of glucocortoids. Metabolism, 23, 159-202.

Weiss, S. J., Lampert, M. B. and Test, S. T. 1983. Long-lived oxidants generated by human neutrophils: Characterization and bioactivity. Science, 222, 625-626.

van der Zijpp, A. J. 1983. The effect of genetic origin, source of antigen, and dose of antigen on the immune response of cockerels. Poultry Sci., 62, 205-211.

INTRODUCTION

It is generally accepted that severe chronic stress strongly enhances the occurrence of organ damage, like stomach ulcers and cardiovascular diseases. This has been shown in animals and men (for a recent review see Meerson, 1984). Such forms of damage are not uncommon in farm animals as illustrated in the contribution of Dämmrich. Under stress conditions pigs and calves may show stomach wall damage and, although the type or the amount of food ingested may play a significant role in the etiology of these somatic aberrations, the question remains as to how far in these cases the stress state as such may be a crucial factor. A similar point holds for the myopathological changes observed in horses and especially in pigs as described by Dämmrich. Again cardiovascular capability and blood circulation in the muscles are crucial in animals that are the product of a selection history that favoured muscle growth. However, it is not unlikely that stress as such may strongly contribute to the observed abnormalities. The mechanisms underlying such possible processes have been described by Bohus et al., who show how a decrease in predictability and/or controllability of the environment may change autonomic nervous activities and by this, for instance, characteristics of the cardiovascular system.

It becomes more and more clear that for a full understanding of stress we not only should know which causal factors function as stressors, but also that we should develop a more integral view on the whole stress state. This view implies Selye's General Adaptation Syndrome, Mason's pattern of endocrine responses and, last but not least, the action pattern of the autonomic nervous system. Since Cannon we know about the significance of the sympathetic system in the context of alarm and stress. It is very relevant that individual states of stress also express the active involvement of the parasympathetic system, as indicated by Bohus et al. These authors refer to interesting findings that show that stress as such not necessarily implies, tachycardia, for instance; also bradycardia may occur, reflecting the active involvement of the parasympathetic system. Bohus et al. accept stress as a state of the organism resulting from a decreased certainty about environmental changes, which is basically a cognitive deficit. This point will be discussed in more detail in the following chapters. The idea that cognitive deficits may significantly contribute to organ damage, as described by Dämmrich, has hardly been explored. It seems a worthwhile one, because of its direct connection with (still) vague concepts like animal well-being.

In the work of Bohus et al. emphasis is put on the finding of types of coping. Some animals behave actively when being overtaxed, others passively. These differences appear to be individual characteristics in the sense that one and the same individual tends to behave actively or passively but not both ways. Such individual constraints in coping are very interesting and summon us to focus on individuals (and not on the abstract average animal), when we want to explain the way animals (and men) cope with stressors. It further has great significance for all considerations about farm animal welfare, since it underlines the idea that we never can evaluate the individual state of an organism by measuring its behaviour only, or endocrinological (re)actions only, etc. A complete and reliable picture can only be realized if we are able to describe a more integrated picture of what the animal does behaviourally and physiologically. Moreover we never should forget that animals are biological systems that change over time as a result of experience.

REFERENCES

Meerson, F.L., 1984. Adaptation, stress and prophylaxis. Springer, Berlin (translated from Russian).

PHYSIOLOGY OF STRESS: A BEHAVIORAL VIEW

B. Bohus, J.M. Koolhaas, C. Nyakas, A.B. Steffens,
D.S. Fokkema, A.J.W. Scheurink

Department of Animal Physiology,
University of Groningen,
Biological Center, P.O. Box 14,
9750 AA Haren, The Netherlands.

ABSTRACT

Stress is viewed as a general biological and usually functional response to environmental and bodily demands. This paper summarises recent findings on interactions between environment, individual behavioral and physiological characteristics and the properties of stressors, stress states (responses) and organ systems in determining health or disease.

INTRODUCTION

Adaptation to demands of social and physical environment by means of behavioral, neuroendocrine, autonomic physiological and metabolic responses in order to maintain homeostasis is a general biological phenomenon in animal and man. Cannon (1915) was the first to formulate a physiological concept of adaptation emphasizing the necessity of links between brain, behavior and the endocrine system. Cannon's (1915) emergency reaction involved the catecholamine systems that serve to rapidly mobilize resources for the metabolic requirements of fight or flight. The term stress as introduced by Selye (1935) emphasized the role of the adrenal cortex as a major organizer of non-specific bodily responses to environmental demands. The operationalization of this concept led to the idea that the level of corticosteroids in the blood reflects whether an organism is in a stress state (heightened physiological reactions), and whether a condition (stressor) is demanding on the organism. Selye's (1950) "General Adaptation Syndrome" hypothesis presupposed an alarm, a resistance and an exhaustion phase of stress reactions of the adrenal cortex. These three phases could represent a continuum from adaption (health) or maladaption (disease or death). Behaviorally oriented physiological and endocrine research in animal and man resulted in a marked change in our view on environment, stress and adaptation. However, the operational definition of stress based upon plasma corticosteroid levels has surprisingly remained the major theoretical frame for many physiologists and endocrinologists. This theoretical frame does not consider the brain and behavior as important variables of the stress state.

Mason (1971) has introduced the importance of behavioral factors in pituitary-adrenal (and other endocrine system) responses to stressors, and thereby the idea of a certain specificity in stress states. The analysis of multiple hormonal system responses under real life conditions showed that the complete pattern of neuroendocrine reaction may be highly specific for certain environmental conditions. Subsequent developments led to the recognition of the importance of such environmental characteristics such as controllability, predictability, certainty, uncertainty and that of coping behavior in stress physiology and pathology (see Seligman and Weiss, 1980).

The present paper describes a behavioral view on the physiology and pathology of stress. A novel, behavioral physiological stress concept as presented here takes its origin from a Selyean view - i.e. stress is a response - but it is extended to environment, behavior and physiology. It also incorporates up-to-date neuroendocrine knowledge on the brain as a major endocrine organ both as a source and a target of the hormones. In addition, this concept is based on the recognition of interactions between the various elements of the stressors and stress. Accordingly, the classical view that stress can be considered as a collection of more or less reflex responses is discarded.

In the recent concept, stress is viewed as a general biological and usually functional response to environmental and bodily demands. Whether stress is a challenge or a threat - health or disease - depends upon interactions between (1) environment, individual characteristics and the properties of stressors, stress and the physiological systems, and (2) the nervous system, peripheral organ systems and the neuroendocrine system. Within these latter interactions four levels of organization of a stress reaction are apparent: the limbic-midbrain system, the hypothalamus, the pituitary gland-peripheral endocrine organs and the target organs. The target organs include physiological systems (e.g. cardiovascular system, GI tract, liver etc.), but also the brain being the major target organ from an organizational point of view. The means of communication between the first 3 levels and the targets are the autonomic nervous system, neuroendocrine principles and neurally active substances originating from the GI tract, immune system, etc. (Bohus, 1984a). The emphasis in this paper is placed upon the interaction of factors listed under number one. The validity of this view is discussed on the basis of combined behavioral, neurophysiological, physiological and neuroendocrine studies as performed in the laboratory rat in our department during the last few years. The main emphasis is given to the cardiovascular (blood pressure and cardiac rhythm), respiratory, metabolic and neuroendocrine reactions.

ENVIRONMENT, STRESSOR, COPING, RESPONSE AND SYSTEM: THE INTERACTING FACTORS

The various elements of the interacting factors that contribute to an ultimate stress state of health and disease are depicted in Table 1. Our behavioral physiological stress concept emphasizes that the interaction between the major factors is of primary importance. Therefore, the following separate discussion of the characteristics of the various elements is obviously arbitrary. Practically all phenomena that will be described here could be listed under all headings.

TABLE 1 INTERACTING FACTORS DETERMINING STRESS-RELATED HEALTH OR DISEASE

STRESSOR		COPING	
Kind	: Psychosocial Physical Psysico-chemical Internal	Kind	: Active Passive
Duration/Frequency:	Short Long Incidental Regular	Origin:	Genetic Perinatal experience Premature experience Adult experience Age Sex
	ENVIRONMENT Controllable (Loss of Control) Predictable (Unpredictable)		
RESPONSE		SYSTEM	
Quality :	Specific Non-specific	Kind :	Behavioral Endocrine Physiological
Quantity:	Magnitude Duration	State:	Baseline Stress Recovery Physiological Pathological

The stressor: physiological and neuroendocrine responses to psychological (emotional), social and physical stressors

It is a classical view that is maintained in study- and textbooks that a typical stress reaction pattern is a mass-discharge of the sympathetic nervous system. The mass-discharge is followed by the activation of sympathetically facilitated (e.g. cardiovascular system) and inhibited (e.g. endocrine pancreas, GI-tract, etc.) system functions. There are a number of studies suggesting that this is not necessarily the case.

A relative bradycardiac heart response to an emotional stressor in conjunction with blood pressure increase has been observed in young adult Wistar rats in an inhibitory avoidance situation (Bohus, 1974, 1985; Van der Meulen and Bohus, 1984). Since baroreceptor reflex bradycardia as a consequence of blood pressure elevation can be excluded as the cause of the heart rate slowing (Bohus et al., 1976), a parallel activation of parasympathetic and sympathetic nervous systems may be responsible for the differential cardiovascular changes in this situation.

Vagal overactivity is sometimes considered as a metabolically irrelevant response in stressful situations (Obrist, 1981). However, it may serve to temper blood pressure changes as induced by certain emotional stimuli. Immobility is often the major behavioral stress reaction in situations where a passive behavioral strategy is applied. The immobile form of natural defense behavior in cats is accompanied by bradycardia and a decrease in cardiac output (Adams et al., 1958, 1971) and only a minimal blood pressure rise occurs (Mancia et al., 1971) despite the fact that the animal is "stressed". A number of findings suggests that the sympathetic and parasympathetic system can also act non-reciprocally. This 'non-reciprocity' is organized either at hypothalamic (Kollai and Koizumi, 1985) or brainstem levels (Langhorst et al., 1981).

Sex, age and perinatal experiences are superimposed factors on the bradycardiac heart reaction to emotional stressors in the inhibitory avoidance situation. There is a sexual dimorphism in the cardiac response in young adult Wistar rat (De Loos et al., 1979; Bohus, 1985). Females in estrus fail to show a bradycardiac response to the emotional stressor. There is a bradycardiac response during the diestrus phase of the estrus cycle, but its magnitude is less than in the males. Experiments with castration and hormone replacement in males and females suggest that estrogens are primarily responsible for the dimorphism. Aging results in a diminished magnitude of the bradycardiac response of male Wistar rats (Nyakas et al.,

1986). This diminution is probably a consequence of a decreased parasympathetic drive from the central nervous system. Finally, neonatal stress of handling is followed by a diminished bradycardiac response to emotional stress in adulthood (Bohus and Cottrell, 1985). This finding agrees with the reports by Denenberg (1964), Ader (1968) and others showing diminished behavioral and endocrine stress reactions in adulthood following neonatal stress.

A different response pattern is emerging if male rats of the TMD S3 strain are exposed to social stimuli like an intruder in their own territory or to a dominant encounter in a neutral environment. Blood pressure, heart rate, plasma catecholamine and corticosterone levels rise rapidly. There is, however, a clear difference whether the animal is victorious or defeated. The mean responses are much larger in magnitude and/or duration as the consequence of defeat (Schuurman, 1980; Fokkema and Koolhaas, 1985). The pattern is changing if the rats are exposed repeatedly to the dominant rat but without actual physical contact (encaged dominant). This situation, called a psychosocial stressor, leads to a hypertensive response but heart rate is decreasing (Fokkema and Koolhaas, 1985; Koolhaas et al., unpublished). Thus, social stressors evoking fight or flight result in a sympathetic hyperactivity. Psychosocial stressors induce mostly immobility behavior and, alike in the case of inhibitory avoidance, there is a parallel activation of certain parasympathetic and sympathetic mechanisms.

Another response pattern is seen in male Wistar rats when subjected to work-load of forced swimming (Scheurink and Steffens, in prep.). Blood pressure, heart rate, plasma epinephrine (E) and norepinephrine (NE), free fatty acids (FFA), and blood glucose levels increase differentially before, during and after repeated swimming. Already before the onset of the 15 min. swimming period blood pressure, heart rate and plasma E levels are increasing. After about 6-8 min. swimming these measures show a considerable fall. At the same time plasma NE, FFA and blood glucose levels begin to rise and plasma insulin level decreases. The NE, FFA and glucose levels reach their maximum at about the end of the 15-min. swimming period. The post-swimming period is characterized by a secondary increase in blood pressure, heart rate and plasma E levels. Plasma NE, FFA and blood glucose levels gradually decline in this period that is mostly accompanied by wet-shaking and vigorous grooming behavior.

This pattern seemingly fits into the picture of classical sympathetic stress-discharge, but the differential nature of the patterns suggest a highly specifically organized response. The first phase of the response may be the result of prediction of circulatory needs of an expected stress state, which is then tuned in accordance with the demand. The second phase is related to metabolic needs. This complex pattern may be the consequence of differential activation of the various subdivisions of the sympathetic nervous system.

Individual characteristics

Both from a physiological and a pathological point of view the attention has long been focussed on individual differences in stress reactions and on the correlations between the various behavioral, physiological and endocrine responses (e.g. Lissák and Endröczi, 1960; Bohus et al., 1963; Henry, 1976; Ladewig, this volume). During the last few years the results of a large number of studies in our laboratory suggested that, on the basis of the variation of the kind and the magnitude of behavioral responses to various environments in rats and mice, it is possible to distinguish active and passive behavioral strategies (Koolhaas et al., 1986). Active rats and mice seem to have high demand to control their environment, they use a large number of routines to accomplish the control and they are less dependent on actual environmental stimuli. In contrast, passive animals show low demand for control or they are ambivalent; their behavior is markedly directed by actual environmental signals.

Individual variation is a superimposed factor on the behavioral, physiological and neuroendocrine response patterns to psychological and social stressors as described previously. For example, the magnitude of the bradycardiac response of young adult male Wistar rats to an emotional stressor and measures of the exploratory behavior of the same animals in a novel environment are negatively correlated. Rats that explore more, show less bradycardia, while the reverse holds for less exploring animals (Nyakas et al., 1986). Correlations have also been found between the bradycardiac reaction and the immobility behavior in response to the emotional stressor: the longer the period of immobility the more marked bradycardiac stress reactions is seen (De Loos et al., in prep.).

Fokkema and Koolhaas (1985) found positive correlations between aggressive behavior, blood pressure and plasma catecholamine levels in social victory and defeat situations. The more aggressive (offensive) rats show higher blood pressure response, and higher neuroendocrine (plasma epinephrine, norepinephrine and corticosterone) responsiveness or ratios.

These and other studies suggest that active rats (more aggressive offensive-defensive behavior, more exploration, less immobility, etc.) show physiological and neuroendocrine responses that reflect sympathetic activity. In contrast, passive rats (less or no aggressive behavior, less exploration, more immobility, etc.) exhibit less sympathetic and more parasympathetic dominated physiological activities. Since these individual characteristics are reflected in behavior, physiology and the neuroendocrine system and manifest themselves across diverse biotic and non-biotic situations, it seems to be justified to distinguish between active and passive coping strategies. It is necessary to mention, that Henry's suggestion to distinguish the Cannonian fight-or-flight kind of behavioral and endocrine characteristics and the Selyean distress (conservation-withdrawal) type of reactivities of laboratory mice living in colonies (see Henry and Stephens-Larson, 1985), is rather comparable to the active-passive dimension of coping strategy as advocated by us.

Response specificity and environmental variables

In the foregoing sections we have described a few response patterns to (psychological, psychosocial and physical) stressors that showed particular temporal or directional characteristics. Here we want to summarize findings suggesting that the appearance of certain physiological stress reactions depend on environmental factors such as predictability and controllability.

Fokkema et al. (1986) observed that episodes of a large amplitude blood pressure oscillations occur in the defeated rat during territorial fight. A similar phenomenon could be observed during so-called psychosocial stress (confrontations with the encaged dominant). The blood pressure oscillation coincides with the respiratory pattern known as pressure breathing. The intrathoracic pressure is strongly positive and prolonged expiration is observed while the blood pressure rises and heart rate decreases. These episodes outlast the actual fight and never occur in the victorious rats. Episodes of large amplitude blood pressure oscillations with heart rate decrease have been infrequently observed in Wistar rats during immobility behavior in the inhibitory avoidance situation (Bohus et al., in prep.). It is conceivable that the occurrence of this phenomenon is not simply a question of quantitative properties of the stressor, but rather of qualitative characteristics of the environment. The defeated rat may be in a peculiar stress state, because of unpredictable consequences of the meeting with an aggressive encounter. This state seems to be preserved without the presence of the actual stimulus once it was induced. The occurrence of this pattern may also depend on the environmental factor of controllability - loss of control. The defeated rat lost control of the social situation.

The physiological or pathological importance of pressure breathing and consequent blood pressure fluctuation is not clear yet. Rapid alterations in blood pressure may have consequences for the development of cardiovascular pathology as suggested by some human studies (Grossman, 1983).

Long-term consequences of short-term stress

Recent epidemiological studies in man suggest that exposure to short-term stressors may result in long-term alterations in mental and somatic functions, and thereby contribute to the development of disease (e.g. Siegrist, 1984). Preliminary observations in our laboratory show that such a phenomenon may occur in animals too (Koolhaas et al., in prep.).

Male TMD-S3 rats defeated by a dominant in a neutral environment (social stressor of defeat) exhibit a marked change in their daily food intake pattern for a few weeks. Normally the food intake of the rat is much higher during the night period. This circadian pattern is typical for nocturnal mammals. Following defeat the circadian rhythmicity disappears: the meal periods are randomly scattered over the whole 24 hr. An important factor of this stress effect is again the kind of experience the rat received. Defeat changes the food intake pattern provided that the animals experienced victory before. The physiological and/or pathological background of this long-term stress reaction remains to be studied. The fact that body weight gain of the animals remains behind their controls suggests that some kind of temporary anorexia develops. As an environmental factor the importance of change in predictability may be implicated.

Long-term social stressor: cardiovascular and immunological consequences

The rat is a social creature and lives in groups under natural conditions. There is, however, a marked hierarchy within the male population with one dominant animal and a group of rats that are subordinates or submissive to the dominant. Some previous investigations (e.g. Alexander, 1974; Henry and Stephens, 1977; Harrap et al., 1984) suggested that long-term social stress in rat and mice as studied in colony conditions in the laboratory results in pathological alterations such as hypertension, arteriosclerosis, gastric ulcer, etc. Fokkema (1985) raised the question as to whether behavioral and physiological characteristics of the rats (i.e. active or passive coping strategies) represent a risk factor for cardiovascular disease provided that the animals are exposed to long-term social stressors in a laboratory colony situation. It appeared that under the given conditions a group of subdominant males rose in the hierarchy inbetween the single dominant and the submissives (subordinates). These subdominants have been characterized before the colony aggregation as active animals on the basis of their territorial offensive and defensive/flight behavior. These rats are regularly challenging the position of the dominant, and thereby they receive the most offensive behavioral stimuli from the dominant. The submissives originate from rats displaying passive strategy in dyadic interactions. Because of their continuous passive (freezing) strategy they seem to receive few stimulation in the colony.

Following a 3-months period in the colony the mean blood pressure of the subdominant rats became significantly higher than that of the dominant or of the submissives. The subdominant rats, that have higher blood pressure, show also a high plasma catecholamine response to social stimulation. This higher reactivity may contribute to the elevation of blood pressure of the subdominant rats.

Recent observations in cooperation with Ballieux and Heijnen (Koolhaas et al., in prep.) suggest that long-term social experience has also differential effect on the immune system of the rat. Following a 3-months colony aggregation there is a suppression of total blood lymphocytes in the subdominants (in comparison to the dominant and the submissives). A higher T-helper/T-cytotoxic-suppressor ratio, a higher interleukine$_2$ production, a strong thymus involution, but a normal proliferation response in the spleen to CON-A stimulation have been found in the subdominant rats (cf. Ballieux and Heijnen, this volume). In contrast, subordinate rats show a very low proliferation response.

Taken together, long-term social stress as a consequence of colony aggregation affectsphysiological systems like the cardiovascular and the immune ones. The alterations cannot be considered yet as pathological ones, but they may represent a risk for the development of disease. The major finding of these experiments is, that the alterations are rather differential; one population of rats, the subdominants, is at the most risk. Since the position in the colony is clearly a function of the basic behavioral characteristics of the animals, active behavioral and physiological response strategy is a potential risk for disease. Active rats demand control of their (social) environment. Subdominants' demand to control in the colony is suppressed. Accordingly, the observed organ system changes are the result of interactions between the kind of coping with the environment (uncontrollable or loss of control), the duration of the stressor, and the given physiological system.

Pathological state of the system: hypertension and physiological responses

Another important variable of the reactions to stressors is the state of the given organ system. Recent observations show, that the magnitude and the duration of the cardiac response to an emotional stressor in the inhibitory avoidance test, is dependent on the systolic blood pressure level of the rats (Nyakas and Bohus, in prep.). It was found that genetic (spontaneous hypertensive rats, SHR), renal and deoxycorticosterone/salt (DOCA/salt) hypertensive rats show a bradycardiac response alike normotensive Wistar controls, provided that their systolic blood pressure under resting conditions does not exceed 170 mm Hg. Rats having a systolic blood pressure higher than this value display a tachycardiac heart reaction. Accordingly, the risk of cardiac dysfunction such as ventricular fibrillation, etc. in reaction to an emotional stressor is markedly increased under pathological conditions of the cardiovascular system.

Fokkema and Koolhaas (1985) found that the mean blood pressure of TMD-S3 rats that lost a fight in a dyadic interaction is slightly but significantly lower the day after the defeat as compared to pre-defeat values. Victory experience fails to have a similar action. Recent findings suggest that loss of control in a social or non-social situation (defeat or forced swimming) results in a relatively long lasting change in blood pressure and cardiac reaction to an emotional stressor in hypertensive rats (Nyakas and Bohus, in prep.). SHR or DOCA/salt hypertensive rats show a very substantial fall in resting systolic blood pressure (± 25 mm Hg) for a couple of days following a repeated defeat or forced swimming. Normotensive control rats fail to show a similar response to the stress of defeat or forced swimming.

The cardiovascular state of the hypertensive rats with decreased blood pressure suggests a shift towards baroreceptor regulation. Blood pressure decrease is accompanied by a markedly increased heart rate. The cardiac response to the emotional stressor of fear from painful stimulus is also changed. The bradycardiac response is diminished. The vagally mediated cardiac response of the normotensive stressed controls remains unchanged. Accordingly, diverse stressors differently affect the organ system reactions under normotensive and hypertensive conditions. It is likely that the risk of failure to adapt to the environment is markedly increased under a pathological condition such as high blood pressure.

GENERAL DISCUSSION

A behavioral physiological view on stress-related processes is based upon the recognition that stress is a multifactorial reaction, and the various factors such as environment, coping, stressor, response and system highly interact in determining the final stress state of the organisms. The data from our laboratory as well as the findings of others (e.g. Henry and Stephens-Larson, 1985; Dantzer and Mormède, 1983) suggest, that measuring a single endocrine or physiological parameter cannot give an objective view of the stress state of the organism. From a physiological point of view, stress is a healthy reaction of the organism to cope with the environmental and bodily demands, sometimes under extreme conditions.

It isn't easy to determine the border between physiology and pathology. Certain individual characteristics represent a risk for certain diseases. The involvement of personality factors in the development and outcome of diseases in man is a long-standing question which is not well answered yet. Recent findings (Koolhaas et al., 1983, 1986) suggest that one should face the question of 'personality' in animals as well. Since the dimension controllability-loss of control has an important impact on the development of behavioral stereotypies in farm animals (see Wiepkema, in this volume) it will be of interest to investigate whether individual characteristics represent a major factor in stress-related stereotyped behaviors.

Our studies also emphasized the potential role of controllability and predictability in determining specific features of stress reaction. This is in agreement with the studies of Weiss (1968, 1970), Bassett and Cairncross (1976), Sklar and Anisman (1979) and others who showed that loss of control and/or unpredictability are major environmental factors involved in stress-related diseases. The recent studies add two more dimensions to these factors: individual characteristics, and controllability and predictability in seminaturel social settings.

Little is known yet about the pathomechanism of stress-related diseases. Dependence on the organ system, dependence on sustained activation, dependence on recovery after stress, dissociations between behavior and physiology, the brain and the periphery (e.g. reflex regulation instead of adaptive higher nervous control) may be important factors of maladaptation and disease. However, it is clear that not a single one of these factors alone can cause diseases. It is a challenge to investigate the pathophysiological mechanism from molecular upto organismal level in order to determine the border between physiological and pathological stress states. Neuroendocrine mechanisms - i.e. hormonal states - both as an integral component of stress reaction and a major modulator of integrative central nervous process may have a keyfunction in health or disease (Bohus 1984b).

ACKNOWLEDGEMENTS

These results were partly supported by the Dutch Heart Foundation (project nr. 80.019 (D.S.F.) and nr. 84.002 (C.N.) and the Foundation for Biological Research BION (nr. 427.102 (A.J.W.S.). C.N. is on leave of absence from the Department of Experimental and Clinical Laboratory Investigations, Postgraduate Medical School, Budapest, Hungary.
The authors wish to thank Mrs. Joke Poelstra-Hiddinga and Ms. Claudia Krol for their editorial assistance.

REFERENCES

Adams, D.B., Baccelli, G., Mancia, G. and Zanchetti, A. 1968. Cardiovascular changes during preparation for fighting behavior in the cat. Nature 220: 1239-1240.

Adams, D.B., Baccelli, G., Mancia, G. and Zancetti, A. 1971. Relation of cardiovascular changes in fighting to emotion and exercise. J. Physiol. 212: 231-328.

Ader, R. 1968. Effects of early experiences on emotional and physiological reactivity in the rat. J. comp. physiol. Psychol., 66: 264-268.

Alexander, N. 1974. Psychosocial hypertension in members of a Wistar rat colony. Proc. Soc. Exp. Biol. Med. 146: 163-169.

Bassett, J.R. and Cairncross, K.D. 1976. Myocardial sensitivity to catecholamines following exposures of rats to irregular, signalled footshock. Pharmacol. Biochem. Behav. 4: 27-37.

Bohus, B. 1984a. Neuroendocrine interactions with brain and behavior: A model for psychoneuroimmunology? In: "Breakdown in Human Adaptation to Stress. Towards a Multidisciplinary Approach". (Ed. R.E. Ballieux) (Martinus Nijhoff Publ., The Hague). pp. 638-652.

Bohus, B., 1984b. Endocrine influence on disease outcome: experimental findings and implications. J. Psychosom. Res. 28: 429-438.

Bohus, B. 1985. Acute cardiac responses to emotional stressors in the rat: the involvement of neuroendocrine mechanisms. in: "Psychophysiology of cardiovascular control. Models, Methods and Data". (Eds. J.F. Orlebeke, G. Mulder and L.J.P. van Doornen), (Plenum Press, New York). pp. 131-150.

Bohus, B. and Cottrell, G. 1985. Neuro-peptides and sex hormones: effects of emotional behavior and cardiac responses. In: "Paediatric Neuroendocrinology". (Ed. D. Gupta). (Croon Helm, London). pp. 20-37.

Bohus, B., De Jong, W., Provoost, A.P. and De Wied, D. 1976. Emotionales Verhalten und Reaktionen des Kreislaufs und Endokriniums bei Ratten. In: "Seelische und körperliche Störungen durch Stress". (Ed. A.W. von Eiff). (Gustav Fischer Verlag, Stuttgart). pp. 140-157.
Bohus, B., Endröczi, E. and Lissák, K. 1963. Correlations between avoiding conditioned reflex activity and pituitary-adrenocortical function in the rat. Acta Physiol. Sci. hung. 24: 79-83.
Cannon, W.B. 1915. Bodily Changes in Pain, Hunger, Fear and Rage. (Appleton. New York).
Dantzer, R. and Mormède, P. 1983. Stress in farm animals: a need for reevaluation. J. Anim. Sci. 57: 6-18.
De Loos, W.S., Bohus, B., De Jong, W. and De Wied, D. 1979. Reduction of heart-rate reactions to emotional stress by ovarian hormones in rats. J. Endocrinol. 81: 138p-139p.
Denenberg, V. 1964. Critical periods, stimulus input and emotional reactivity: a theory of infantile stimulation. Psychol. Rev. 71: 335-365.
Fokkema, D.S. 1985. Social behavior and blood pressure: a study in the rat. Ph.D. Thesis. University of Groningen.
Grossmann, P. 1873. Respiration, stress and cardiovascular function. Psychophysiology 20: 284-300.
Harrap, S.B., Louis, W.J. and Doyle, A.E. 1984. Failure of psychosocial stress to induce chronic hypertension in the rat. J. Hypertension 2: 653-662.
Henry, J.P. 1976. Mechanisms of psychosomatic disease in animals. Adv. in Vet. Sci. 20: 115-145.
Henry, J.P. and Stephens, P.N. 1977. Stress, Health and the Social Environment. A Sociobiological Approach to Medicine. (Springer, New York).
Henry, J.P. and Stephens-Larson, P. 1985. Specific effects on stress in disease processes. In: "Animal Stress". (Ed. G.P. Moberg). (American Physiological Society, Bethesda, Maryland). pp. 161-175.
Kollai, M. and Koizumi, K. 1981. Cardiovascular reflexes and interrelationships between sympathetic and parasympathetic activity. J. Auton. Nerv. Syst. 4: 135-148.
Koolhaas, J.M., Schuurman, T. and Fokkema, D.S. 1983. Social behavior of rats as a model for the psychophysiology of hypertension. In: "Biobehavioral Bases of Coronary Heart Disease". (Eds. T.M. Dembroski, T.H. Schmidt and G. Blümchen). Vol. 2. (Karger, Basel). pp. 391-400.
Koolhaas, J.M., Fokkema, D.S., Bohus, B. and Van Oortmerssen, G.A. 1986. Individual differentiation in blood pressure reactivity and behavior of male rats. In: "Biobehavioral Bases of Coronary Heart Disease". (Eds. T.M. Dembroski, T.H. Schmidt and G. Blümchen). Vol. 3. (Karger, Basel). In press.
Langhorst, P., Lambertz, M., and Schulz, G. 1981. Central control of interaction affecting sympathetic and parasympathetic activity. J. Auton. Nerv. Syst. 4: 149-163.
Lissák, K. and Endröczi, E. 1960. "Die Neuroendokriene Steuerung der Adaptionstätigkeit". (Akademische Verlag, Budapest). pp. 172.
Mancia, G., Baccelli, G. and Zanchetti, A. 1972. Neurodynamic responses to different emotional stimuli: patterns of mechanims. Am. J. Physiol. 223: 925-933.
Mason, J.W. 1971. A re-evaluation of the concept of 'non-specificity' in stress theory. J. Psychiat. Res. 8: 323-333.
Nyakas, C., Alingh Prins, A.J. and Bohus, B. 1986. Cardiac responses and behavioral reactivity to emotional stress in aged rats. Proc. 27th Dutch Fed. Meeting, Abstr. no. 295.
Obrist, P.A. 1981. Cardiovascular Psychophysiology, A perspective. (Plenum Press, New York).

Schuurman, T. 1980. Hormonal correlations of agonistic behavior in adult male rats. In: "Adaptive capabilities of the nervous system". (Eds. P.S. McConnel, G.J. Boer, H.J. Romijn, N.E. van de Poll and M.A. Corner). Progress in Brain Research, Vol. 53. (Elsevier, Amsterdam). pp. 415-420.
Seligman, M.E.P. and Weiss, J.M. 1980. Coping behavior: learned helplessness, physiological change and learned inactivity. Behav. Res. Theor. 18: 459-512.
Selye, H. 1935. A syndrome produced by diverse nocuous agents. Nature 138: 32-33.
Selye, H. 1950. Stress. The physiology and Pathology of Exposures to Stress. (Acta Medica Publ., Montreal).
Siegrist, J. 1984. Interaction between short- and long-term stress in cardiovascular disease. In: "Breakdown in Human Adaptation to 'Stress'". Vol. 2. (Ed. A. L'Abbate). (Martinus Nijhoff, Boston). pp. 892-899.
Sklar, L.S. and Anisman, H. 1979. Stress and coping factors influence tumor growth. Science 205: 513-515.
Van der Meulen, J. and Bohus, B. 1984. Adrenalectomy prevents behaviorally-induced hypertension in the rat. Neurosci. Lett. Suppl. 18: S376.
Weiss, J.M. 1968. Effects of coping on stress. J. comp. physiol. Psychol. 65: 251-260.
Weiss, J.M. 1970. Somatic effects of predictable and unpredictable shock. Psychosom. Med. 32: 397-408.

ORGAN CHANGE AND DAMAGE DURING STRESS - MORPHOLOGICAL DIAGNOSIS -

Klaus Dämmrich
Institut für Veterinär-Pathologie
Freie Universität Berlin
Strasse 518 Nr. 15, D 1000 Berlin 37

ABSTRACT

In pathology, stress research comprises the influence and action of stressors on organ structures. If the influence of stressors is in balance with the organism's predisposition the stressor is compensated by the adaptability of the organism and stress reactions develop. The morphological diagnosis is based upon the differentiation of two types of reactions. First the nonspecific reaction with the general adaptation syndrom (GAS), and secondly the specific reaction caused by the direct attack of the stressor, with the possibility of lower resistance of organs against stressors. The different kinds of stress reactions are demonstrated in the problem of gastric and pyloric ulcers in pigs and calves, with the paralytic myoglobinuria in horses, and with the porcine stress syndrome.

INTRODUCTION

The morphological aspects of "stress" and the relationship between "stress" and organ lesions are presented in two parts. The first part includes more general aspects, while in the second one concrete examples in domestic animals are described. The whole life is determined by stressors. Under physiological conditions stressors promote the various systems of adaptation from which the survival of the organisms depends. Under conditions without stressors animals cannot survive, or in other words "stressors are the fire of life" (SELYE). Physiological adaptability depends on the balance between stressor and predisposition. Predisposition is defined as the sum of constitution and condition. Mainly hereditary factors determine the constitution and acquired properties characterize the condition of animals. Lower grades of imbalance are compensated by the adaptability of the organism. This adaptability guarantees health or fitness of animals. Integrity of constitution and condition is a precondition for proper adaptability. Interventions in constitution or condition disturb adaptability, which may result in malfunction or organ lesions. The term "stress" characterizes situations, in which the integrity of the organism is injured by imbalances between predisposition and stressors (table 1).

Table 1:

	IMBALANCE	
predisposition	------------------	Stressor
lower resistance		excessive
	overtaxed adaptability	
specific reactions		nonspecific reaction
organ lesions		general adaptation syndrome

Imbalances can be caused by excessive stressors or by lower resistance against stressors. The lower resistance depends from regressive changes of constitution or condition. The imbalance overtaxes the adaptability of the organism.

STRESS REACTIONS:

The nonspecific reaction includes mainly the general adaptation syndrome (GAS, SELYE), and is determined by the quantity of the stressors. The specific reactions comprise, among others, malfunction and organ lesions caused by stressors. The ultimate localisation of these lesions in organs or organ systems depends from the direct specific attack of the stressor, or the higher susceptibility of the organism, or combinations of stressor and susceptibility.

The nonspecific reaction:

Under physiological conditions the GAS enhances the organism's survival by compensating light and moderate stressors through the mechanisms of the alarm reaction. This reaction is finished, when the influence of short acting stressors is over. Long acting stressors, however, lead to the phase of increased resistance.

The morphological control of the alarm reaction and stage of increased resistance results from the activation of a system, which is made up of hypothalamus, adenohypophysis and adrenal cortex. During the alarm reaction hypophyseal adrenocorticotropic hormone stimulates the adrenal cortex and the store of corticosteroids, especially glucocorticosteroids, is secreted. Morphologically the alarm reaction is characterized by a decreased content of lipids in the adrenal cortex. During long acting stressors hypophyseal adrenocorticotropic hormones and corticosteroids are secreted continuously. The stage of increased resistance is characterized by signs of functional hypertrophy and hyperplasia of the secreting cells. An increased number of enlarged basophilic cells appears in the adenohypophysis. The weight of the adrenals is increased, because of the proliferation of cortical cells and as a result of increased lipid contents of the same cortex.

Pathological conditions may also be derived from the GAS. Prolonged influence of stressors under unfavorable environmental conditions can change the regulative phase of increased resistance into a phase of autonomous hypercorticism. Characteristic signs of this phase are hyperplasia of hypophyseal basophilic cells and hyperplasia of the adrenal cortex. The continuous hypersecretion of glucocorticosteroids injures the condition of the organism. Mainly metabolic disturbances are observed. Characteristic symptoms are the katabolic metabolism of proteins, enhanced deposition of body fat and hyperglycemia. In connection with the katabolic protein metabolism an increased infective risk exists, because the humoral and cellular immunity is decreased.

Secondly exhaustion of the hypophyseal adrenal system occurs. Long acting moderate stressors, excessive stressors, or added moderate stressors are the cause of exhaustion. Signs of exhaustion are found in the adrenal cortex, which shows depletion of lipids, necrosis, and hemorrhages. Acute adrenal insufficiency is in such cases the cause of death. Especially young growing animals, which are ill because of a severe enterogenic septicemia, die of acute adrenal insufficiency.

The specific reactions:

If the organism's adaptability is overtaxed, the organism makes an effort to cope by means of the nonspecific stress reaction, called GAS. On the other side the stressor attacks directly the target organ. The specific reaction or the development of organ lesions depends on the intensity and quality of the stressor and on the susceptibility of the target organ and/or of the organism. The specific organ lesions are the more important parts of the stress reaction in the pathology of domesticated animals. Some examples of such lesions may illustrate this.

Gastric ulcers:

Very often gastric ulcers occur in pigs and calves. It is supposed that the ulcers are the result of the nonspecific stress reaction, because in humans severe stressors may evoke the development of gastric ulcers. The pathogenesis of the human "stress ulcer" describes neurohormonal influences, which cause circulatory failures of the final vascular system with ischemia, hyperemia and hypersecretion of the gastric mucosa.

In pigs the ulcers are not localized in the gastric mucosa, but are situated in the pars proventricularis with its less sensitive stratified squamous epithelium. In all probability food is the crucial stressor, because extremely fine food particles occupy and constipate the small fossules of the cutaneous mucosa, from which the necrosis and final ulcers start (Lengnick, 1972; Baustad and Nafstad, 1969).

The incidence of pyloric ulcers in veal calves is very high. Acute chronic ulcers and scars are localized on the pyloric torus or secondary in the pyloric channel. Our investigations led to the following pathogenesis. The pyloric sphincter in young ruminants is more powerful than in older animals. Sphincteric constriction continues till the digestion of milk is finished. In veal calves large volumes (daily 20 litres) of milk substitute or of milk lead to prolonged and strong contractions of the sphincteric muscle. In consequence of this compression the pyloric mucosa becomes anemic. The effect of compression can be increased, if bundles of plant fibers or bezoars are incarcerated in the pyloric channel (table 2).

Table 2: Pathogenesis of pyloric ulcers in veal calves

CAUSAL PATHOGENESIS:

large volume of milk substitute or of milk
increased duration of abomasal digestion
prolonged and strong pyloric constriction

FORMAL PATHOGENESIS:

prolonged anemia and ischemia of the pyloric mucosa

hypoxia &
circulatory disturbances = hypoxic necrosis of villi
= erosion
= ulcer

Compression of the mucosa leads to ischemia in the tissue. Prolonged ischemia causes hypoxic lesions in the pyloric mucosa. The lesions start with circulatory disturbances, like stasis, hyaline microthrombi in widened capillaries, edema, and hemorrhages. They worsen the primary hypoxic lesions, and hypoxic necrosis of villi occurs. Autodigestion changes the focal necrosis to erosions and ulcers. It is concluded, that presumably pyloric ulcers in calves are not the result of the nonspecific stress reaction or GAS, but are caused by the direct attack of the stressor " large volume of milk or milk substitute" (Degen 1982; Groth und Berner, 1971).

Paralytic myoglobinuria:

In horses myopathies are caused by the stressor "intensified locomotion". The occurrence of myopathies depends on the imbalance between predisposition and stressor, but the direct cause of imbalance is a decreased condition. In cases of paralytic myoglobinuria the disease starts in well trained horses after a holiday period of one or several days on full diet and no exercise. During this rest period the muscle fibres accumulate glycogen and this accumulation is the causal agent of the metabolic change from aerobic to anaerobic glycolysis, when enforced locomotion (work of these horses) starts again. The developing lactic acidosis in muscles is combined with a minor blood supply, due to muscle cramp. Lactic acidosis and ischemia are responsible for muscle lesions. The lesions are qualified as hyaline or ZENKER's muscle degeneration with all signs of necrosis (Lindholm et al, 1974; McLean, 1973).

Porcine stress syndrome:

Modern pig breeds are aflicted with several severe diseases, due to imbalance between stressors and predisposition. It is well known, that this imbalance is caused by the lower resistance of constitutional factors and that stressors of relatively low intensity are able to develop stress reactions. One of the specific stress reactions is the porcine stress syndrome (PSS) with myopathies, pale soft exudative (PSE) syndrome of the meat and sudden death (cf. Cassens et al., 1975). The PSS indicates a decreased resistance against stressors and stress susceptibility.

The decreased resistance is the result of selection on muscle mass. The increase of muscle mass leads to changes in muscle structure; these changes are responsible for the lower resistance of the muscle against a stressor like "motoric activity". The causal pathogenesis of myopathies is derived from intramuscular and extramuscular factors.

Intramuscular factor = size of muscle fibres (table 3)

In modern pigs the thickness of muscle fibres is strongly enlarged and especially the type II muscle fibers have changed their size. In German Landrace this type II fibres are strongly enlarged as compared to the same type of fibres in Euopean Wild Boars. Because the distance between capillary and fibre is too long for a sufficient supply with oxygen and an adequate removal of metabolites, this enlargement causes an energy deficit in white fibres with the production and accumulation of lactic acid. The critical size of muscle fibres is near 5000 um^2.

Table 3: Mean size of muscle fibres

M. longissimus dorsi for pigs 200 days old
fibre size (cross section um^2)

	GERMAN LANDRACE	EUROPEAN WILD BOAR
Type I	4995	2030
Type II	7335	2500

Intramuscular factor = metabolic types of muscle fibres:

The type I - fibres are characterized by aerobic glycolysis and type II B - fibres by anaerobic glycolysis with production of lactic acid. The intermediate type II A - fibres work with aerobic and anaerobic glycolysis. In German Landrace the percentage of white muscle fibres has been increased (table 4). This high percentage of white muscle fibres means, that in muscles of German Landrace the production of lactic acid increases and accumulates during motoric activity.

Table 4: Types of muscle fibres.

M. longissmus dorsi / 140 days old pigs
%: types of muscle fibres

	GERM. LANDRACE	EUROP.WILD BOAR
Type I (red)	8	6
Type II A (intermediate)	20	49
Type II B (white)	72	45

Extramuscular factors:

Two facts are important in the PSS: the low size of the heart and, as a result of selective breeding, the enlargement of the muscular periphery (table 5).

Table 5: Relative weight of the chamber myocard

Relative weight of the chamber myocard : % of live body weight
300 days old pigs

GERMAN LANDRACE	EUROPEAN WILD BOAR	HUNGAR. MANGALITZA
(160 kg)	(57 kg)	(107 kg)
0,21 %	0,38 %	0,23 %

In relation to body mass German Landrace and Hungarium Mangalitza have small hearts as compared with the European Wild Boar. For a sufficient blood supply of the muscular periphery during locomotion in German Landrace, the small heart is overstrained, and signs of myocardial hypertrophy develop. In Hungarian Mangalitza the small heart is strained only slightly, because the periphery contains much fatty tissue with a lower blood supply (table 6). Therefore there are no signs of overstrained myocard in Hungarian Mangalitza.

Table 6: Hind quarter: Percentage of skin and subcutaneous fatty tissue.

Hind quarter / 300 days old pigs
% skin and subcutaneous fatty tissue

GERMAN LANDRACE	EUROPEAN WILD BOAR	HUNGARIAN MANGALITZA
28 %	20 %	48 %

Under the influence of the different types of stressors in modern pigs the pathogenesis of myopathies is determined by an energy deficit, which is caused by the thickness of muscle fibres and an insufficient blood supply of the muscular periphery. Energy deficit is accompanied with characteristic morphological and biochemical findings. Morphological signs are hypercontraction of myofibrills, cloudy swelling of mitochondria and tubules, segmental and total necrosis of muscle fibres, and acute back muscle necrosis. Biochemically one finds a decreased content of ATP, an increased exothermic glycogenolysis (hyperthermia), and an increased anaerobic glycolysis with lactic acidosis.

The same condition of energy deficit may evoke the PSE-Syndrome due to stressors, which occur before slaughtering. The sudden death during handling, transport, etc. is the problem of an overstrained small heart with compensatory hypertrophy of the myocardial fibres. The elongated distance between capillary and enlarged fibre makes it difficult to supply sufficiently the myocardial fibres with oxygen. Under the influence of stressors tachycardia with very short diastoles leads to insufficient blood supply of the myocard, and, together with the labile energy metabolism of the thickened fibre, the latent energy deficit becomes manifest. Cloudy and swollen mitochondria are characteristics of myocardial degeneration, which is the cause of sudden death with cardiac insufficiency (Bader, 1982a,b, 1983; Dämmrich, 1978, 1981; Johannsen, 1975; Michel, 1963; Szentkuti und Schlegel, 1985).

The post mortem morphological diagnosis of stress reactions includes nonspecific reactions and specific reactions. The morphological analysis of nonspecific and specific reactions allows to differentiate between specific organ lesions and lesions, which are caused by the nonspecific general adaptation syndrom.

REFERENCES:

Bader, R., 1982a. Histologische Befunde aus licht- und elektronenmikroskopischen Untersuchungen an der Skelettenmuskulatur von gesunden, ausgemästeten Schweinen der Deutschen landrasse.
Zbl. Vet. med. A 29. 458-476.

Bader, R., 1982b. Enzymhistochemische und histometrische Untersuchungen an Skelettenmuskeln von ausgemästeten, gesunden Schweinen der Deutschen Landrasse.
Zb. Vet. med. A 29. 443-457.

Bader, R., 1983. Vergleichende histometrische und histologische Untersuchungen an der Sketlettmuskulatur von Wild- und Hausschweinen.
Berl. Münch. Tierärztl. Wschr. 96. 89-97.

Baustad, B. and Nafstad, I., 1969. Gastric ulcers in swine. Path. vet. 6. 546- 556.

Cassens, R.G., Marple, D.N. and Eikelenboom, G. 1975. Animal physiology and meat quality. In: "Advances in Food Research" (Ed. C.O. Chicester). Vol. 21, 71-155. Acad. Press, New York.

Dämmrich, K., 1978. Wachstumsstörungen bei erhöten Leistungsanforderungen.
Fortschritte Vet. med. Zbl. Vet. med. Suppl. 28. 114-119.

Dämmrich, K., 1981. Exogene und endogene Einflüsse auf Erkrankungen des Bewegungsapparates.
Fortschritte Vet. med. Zbl. Vet. med. Suppl. 32. 140-151.

Degen, B., 1982. Pathologisch-anatomische und -histologische Untersuchungen zur Pathogenese der Pylorusgeschwüre bei Mastkälbern.

Groth, W., und Berner, H., 1971. Untersuchungen über das Labmagengeschwür des Kalbes bei Milchaustauschermast und bei Früentwöhnung. Zbl. vet. med. A. 18. 481-498.

Johannsen, U., 1975. Zum Vorkommen degenerativer Skelettmuskelveränderungen beim Fleischschein.
Mheft. Vet. med. 30. 489-494.

Lengnick, H.D., 1972. Ein Beitrag zur Pathogenese und Ätiologie der Ulcera in der Pars proventricularis des Magens beim

Schwein.
Vet. med. Diss. München.
Lindholm, A., Johannson, H.E. and Kjaersgaard, P., 1974. Acute rhabdomyolysis in standardbred horses.
Acta Vet. Scand. 15. 325-339.
McLean, J.G., 1973. Equine paralytic myoglobinuria (azoturia). A review.
Austr. Vet. J. 40. 41-43.
Michel, G., 1963. Zum Bau der Herzmuskulatur bei Haus- und Wildschweinen.
Zbl. Vet. med. A.10. 389-396.
Szentkuti, L., und Schlegel, O., 1985. Genetische und funktionelle Einflüsse auf Fasertypenanteile und Faserdurchmesser im M. longissimus dorsi und M. semitendinosus von Schweinen. Dt. Tierärztl. Wschr. 92. 93-97.

INTRODUCTION

If stressors evoke stress, in particular because of their psychological implications, such as reducing controllability of environmental changes, then specific nervous activities and processes have to be expected when some function of the organism is overtaxed. Such phenomena concerning nervous activities have indeed been found, as discussed by Warburton. For instance, at the hypothalamic level a significant interaction takes place between noradrenergic and cholinergic systems in the regulation of the CRF release, a hormone that plays a central role in the neuro-endocrine stress response (see Oliverio's contribution). However, these same noradrenergic and cholinergic systems presumably also evoke electrocortical arousal and thereby strongly support the involvement of cognitive processes. In order to obtain an adequate description of and insight into what goes on or what may go on in the brain during stress, i.e. during a state of reduced certainty, we should analyse and interconnect two lines of investigations.

In the first line we should make as explicit as possible which sorts of functions an organism has to possess in order to adapt optimally to the variable environment in which it lives. This living includes aspects like growth, reproduction, rest, behaviour, resistance against predators and pathogens. To optimalize all these aspects, a developing organism should be able to gain information about its environment, to store it and to use it at adequate times and places later on. Such processes not only imply a memory, but also certain claims about the accessibility of such a memory. Further, the organism should also be able to put priorities and to evaluate the effectiveness of its own actions. Briefly, in this first line of investigation we should deal with the question which functions are necessary to design a complex goal seeking system as the intact organism appears to be (cf. the contribution of Toates). Since it is plausible that such functions should be realized somewhere in the central nervous system, we should also explore this system in terms of such functions. This is the second line of investigation.

Although we are still far from a slightly satisfactory point in our understanding of this complex brain-behaviour theme, some highly stimulating data, however, are available. For instance, as indicated by Warburton, it is most interesting that a cortical system like the hippocampus not only appears to be involved in the control of stress steroids, but also in the detection of match (or mismatch) between informationel input and the existing expectancy about this input. The availability of such a match-mismatch mechanism is a prerequisite for each organism that has an interest in predictability/controllability of environmental changes (these concepts are discussed in Wiepkema's chapter). Therefore this putative function of the hippocampus makes this brain structure of great relevance in our neurobiological thinking about stress. Warburton also points to another part of the brain, the amygdala, as being of much significance. The often described adaptation to long lasting stressors presumably results from the activation of some nervous suppressor system, involving the amygdala, and not from some physiological exhaustion.

Although Broom has to conclude that in farm animals relatively little is known about hardware and/or software brain systems operating during stress, he emphasizes that these animals have been designed essentially in the same manner. Quite illustrative are the anticipatory autonomic reactions measured in sheep when a dog approaches. Anticipatory (re)actions deserve much attention in future research, since they may imply a strong emotional component (as put forward by Warburton and others; cf. Simonov, 1986).

It is interesting that, just like in other chapters of this book, individual differences in the neuro-endocrine coping response appear to be of great significance. Warburton demonstrates this point with respect to monkeys, in which social status strongly determines their behavioural and endocrine reactions to chronic and painful stressors. Obviously we cannot permit ourselves to study stress responses in animals, or better individuals, that we know only superficially. Each individual has its own and unique relevant history.

If organisms really have a gating system that selects which information will be processed and which not, Broom's data about a changed responsiveness in sows stalled in a particular way are highly interesting. Do animals like sows have the capability to change their informational input drastically, thereby changing the interpretation of their Umwelt? If so, could sows have means to solve chronic uncertainty by negating it? Which are the cognitive implications of such a "blunting" strategy? The field of neuro-biological aspects of stress presumably contains the central questions of stress. However, we have scarcely entered it.

REFERENCES

Simonov, P.V., 1986. The emotional brain. Plenum Press, New York (translated from Russian).

THE NEUROPSYCHOBIOLOGY OF STRESS RESPONSE CONTROL

D.M. Warburton
Department of Psychology
Reading University
Reading RG 2AL
United Kingdom

ABSTRACT

A wide range of physical stimuli act as stressors and elicit stress responses. However, some stimuli are stressors because of their uncertainty, and the stress response is related to their hippocampal function and the brain cholinergic system that is involved in information processing. Studies of adaptation have implicated the amygdala nuclei in the suppression of responses to chronic stressors. In addition, some events are only interpretable as the result of complex cortical analysis on the basis of the individual's experience and so the cortex must be sending inputs to the hippocampus and the amygdala.

INTRODUCTION

In this chapter, the word "stressors" will be used to describe the agent and the term "stress response" to refer to the reaction of the individual. Distress will be used to refer to the detrimental effects that can occur in the individual and can reduce its fitness. Stressors have the potential for causing distress when the stressor is too intense or too prolonged. In the next Section, the nature of the commonest cause of a stress response is outlined.

STRESSORS

A wide range of physical stimuli, such as pain and heat, act as stressors and elicit stress responses which are a simple monotonic function of the intensity of the physical stimulus. But the stress response to psychological stimuli can be just as large as that produced by physical stressors and the response increases also as a simple monotonic function of the intensity of the stimulus. However, some psychological stimuli are stressors, not merely because of their intensity but because of their distribution in time. Information input produces a much greater stress response if it is presented at unpredictable intervals, even though the density of these stimuli over time is the same (Warburton, 1979). Thus, stressors must be considered in terms of their uncertainty, and the stress response would be inversely proportional to their predictability.

Mason (1975) found that a variety of physical stimuli including heat and cold were not necessarily associated with activation of the adrenocortical system, when it was done so that the animal could not readily detect the change. These pysical manipulations shared one important characteristic; the stressful experience was sudden exposure to novelty. An animal"s response was due to the psychological relevance of stimulus rather than to the particular physical stressors to which they were exposed. Stressors are a consequence of an interaction between the individual animal and the situation.

Situational uncertainty is the result of the many alternative meanings that the input has for the animal, and with the available response choices for that animal, so that certainty must be defined subjectively. For an animal there will be degrees of uncertainty about predictability, i.e. whether an event will happen, what event might occur, when it will happen, and controllability, i.e. what action can be taken.

The important question which must be addressed is how these concepts of predictability and controllability are mediated by brain systems.

CONTROL OF STRESS RESPONSES

Over the last ten years, there has been an increasing number of studies measuring hormones which could be responsive to stressors. The data are most complete for adrenal cortical responses and there is also a significant body of literature measuring urinary catecholamines (adrenalin and noradrenalin(). Other responses include plasma growth hormone, prolactin and testosterone.

Most hormonal systems have some control by hormonal feedback to the hypothalamus. However, there is also regulation by neural pathways that converge on the hypothalamus from other parts of the brain. Since these inputs come from different parts of the brain, it is possible that a large number of synaptic transmitters affect the secretion of hypophysiotropic hormones. Our knowledge of transmitter control is most complete for the control of corticosteroid secretion.

Some of the first studies of stress steroid control gave evidence for a brain noradrenalin pathway inhibiting corticosteroid release. Direct injections of 1-noradrenalin close to the medial hypothalamus inhibited the rise in plasma corticosteroids that was induced by surgical stressors. Similar results were obtained by intraventricular injections of noradrenalin in animals that were stressed by handling (Schiaffini, Motta, Piva and Martini, 1971).

Direct injections of cholinergic compounds into the hypothalamus have suggested that there is a cholinergic system which controls the release of corticotrophin into the blood stream to the pituitary and leads to increased plasma corticosteroids. Krieger and Krieger (1970), found that carbachol, a cholinomimetic, implanted in the median eminence region and mid-mammillary body area, induced an abrupt rise in corticosteroid levels. The effects of carbachol were blocked by means of atropine, a cholinergic blocking agent.

Thus, there is good evidence in favour of noradrenergic inhibition of stress steroid secretion in balance with cholinergic stimulation. (Serotonin may also be an excitatory transmitter but the evidence is not completely convincing). The interaction between the noradrenergic and cholinergic systems appears to be in the hypothalamic region which has been implicated in the release of CRF to stimulate the pituitary.

The involvement of a cholinergic pathway in the control of stress steroid release raises the possibility that steroid control may be related to the cholinergic pathways that control electro-cortical arousal (Warburton, 1981). There are such neurones in the cat which terminate in the median eminence region and have their origins in the brain stem. Endröczi showed that injection of carbachol into the ventral tegmental region in the cat produced both electrocortical arousal and an increase in plasma cortico-steroids (Endröczi et al., 1963a and b). This finding indicates that electrocortical arousal and corticosteroid release are related.

HIPPOCAMPAL CONTROL

Electrocortical arousal is an integral part of the stress response and puts the cortex in a state for action-related information processing (Warburton, 1981). The psychological characteristics of this state are a shift from information storage to information throughput. There is an increase in the selectivity of attention, lowered primary memory capacity, increased selectivity of responses, increased speed of response but lowered accuracy. Altogether, this cognitive stress response is a pattern of changes which have evolved for rapid information processing for action (Warburton and Wesnes, 1984).

Electrocortical arousal is controlled by the cholinergic pathways ascending from the Nucleus basalis of Meynert in the mid-brain. An important part of this system is the hippocampal path-way. The anatomical arrangement of the hippocampus is suitable for a feedback loop to control electrocortical arousal. Novel sensory stimuli in all modalities always induce hippocampal synchroniza-tion of brief duration in the resting animal and cortical desyn-chronization. Stimulation of the hippocampus has been shown to produce changes in attention and electrocortical arousal (see Warburton and Russell, 1969). If the hippocampal formation is part of a feedback loop providing regulation of the activity in the ascending cholinergic arousal pathways, then this control would prevent redundant, i.e. 'expected', stimuli from activating the cortex (Warburton, 1975).

It was suggested some years ago that the hippocampus is involved in the detection of mismatch of information inputs with expectancies (see Pribram, 1967). In addition there are data which show that the hippocampus is involved in controlling the release of stress steroids (Warburton, 1979). Thus, the size of the stress response will depend on the size of the match and mismatch between the animal's expectancies and information input.

As Gray (1982) pointed out, if the entorhinal cortex or its afferents to the dentate gyrus is stimulated, then the response can get gradually bigger (potentiation) or it can get smaller (habituation). It appears that there is some kind of gate in the system (Gray, 1982). Psychologically, this gate is more likely to be opened if there is concomitant activity in the ascending <u>dorsal</u> noradrenergic pathways to the septo-hippocampal system from the brain stem. Activity in these ascending noradrenergic fibres informs the septo-hippocampal system that the information that is reaching it from the entorhinal cortex is biologically significant (Gray, 1982). It is interesting that this noradrenergic pathway is believed to be involved in learning (Warburton, 1983) and could facilitate the consolidation of information, which occurs during adaptation to a stressor.

AMYGDALA CONTROL

Evidence for the involvement of some portion of the amygdala in stress response control has come from studies of adaptation to stressors. Most of our understanding of endocrine responses relates to exposure to acute stressors and, despite its practical importance, there are relatively few studies describing endocrine responses to chronic stressors. Most studies now find that, in both animals and men, there is often a rapid loss of endocrine responsivity to the stimulus, adaptation. This evidence will be considered together with some evidence that systems do not return to their "naive", pre-stressor state.

Adaptive Changes

Mason et al. (1968) assessed the effects of repeated exposure to a stressful experience. Monkeys alternated cycles of 6 h of shock avoidance and 6 h of rest for up to 30 consecutive days after 5 days of training. Adrenal cortical excretion was estimated by urinary 17-hydroxycorticosteroid (17-OHCS). During the first 24 hours of avoidance, the typical pattern of the plasma 17-OHCS levels indicated that, during the first 2 hours, there was an initial, acute evaluation. The size of the response appears to be very large but it did not represent anything like maximal output for the monkey. Then the levels declined, reaching extremely low values by 6 or 12 hours after avoidance onset in some individuals. (The individual differences will be discussed in the next section).

These findings are relevant to Selye's hypothesis that after prolonged exposure to stressors there is a stage of pituitary or adrenal insufficiency referred to as "stage of exhaustion" (Selye, 1937). Given the submaximal values, it seems that these declines in plasma 17-OHCS levels on the first day represent the operation of suppression mechanisms, rather than pituitary or adrenal "exhaustion".

A similar finding was observed for corticosterone and pituitary changes in rats following prolonged exposure to stressors. Pollard et al. (1976) found that, after 20 days of exposure to shock avoidance, animals failed to show any rise in corticosterone; the intense secretory activity observed in the cells of the adenohypophysis that secrete ACTH was no longer observed after 20 days and corticosteroid activity returned to control levels. This lack of pituitary activation following prolonged exposure strongly supports the argument that animals adapt to most stressful stimuli by some suppression mechanism.

This suppression mechanism might be either humoral or neural, or a combination of both. The fact that in some cases profound decreases at 6 or 12 hours may be observed, following rather mild initial rates of rise, while in other animals no comparable secondary decreases are seen following much higher initial elevations, argues against a humoral negative feedback mechanism as a sole explanation for the observed changes.

However, there is evidence for a neural suppression mechanism. A similar temporal pattern of plasma 17-OHCS response occurs during sustained stimulation of the amygdaloid complex in the monkey (Mason et al., 1960; 1961). Partial amygdalectomy consistently reduced but did not necessarily abolish the acute plasma 17-OHCS rise during the first 2 hours of avoidance. Interruption of hippocampal outflow lessened the usual drop in plasma 17-OHCS levels 12 hours after onset of a 6 hour avoidance session in animals subjected to partial amygdalectomy. These findings strongly suggest that amygdala-hippocampal interactions may participate in the integration of the 17-OHCS response during sustained avoidance.

The question of the existence of mechanisms suppressing the pituitary-adrenal cortical activity is also raised by the progressive decline in "basal" urinary 17-OHCS levels as the avoidance sessions were repeated. The 17-OHCS response to avoidance was also markedly diminished in the second and third sessions and the response peak was delayed or shifted to the right.

It is important that, in these subjects, large 17-OHCS increases could be obtained with strong acute shocks or with ACTH infusion in the presence of chronically low 17-OHCS levels. This argues strongly that the hypothalamic-pituitary-adrenal cortical axis still has a substantial response capacity and suggesting that the underlying mechanism is a neural one which is actively suppressing the response.

The shift of the 17-OHCS response peak with repeated experience is interesting in terms of a possible neural mechanism. It is significant that a very similar pattern of delayed as well as diminished 17-OHCS response to avoidance has been observed in monkeys subjected to amygdalectomy (Mason et al. 1960; 1961). Again altered amygdala function is implicated in a neural suppression mechanism.

It is apparent that most stressful psychological stimuli upon re-exposure do not elicit responses of the same size as initially, reflecting adaptation. This phenomenon is analogous to habituation and it is known that the hippocampal-amygdala system is involved in this process.

Non-adaptation of the Systems

Although adaptation occurs for most endocrine responses, the chronically stressed animals are not the same as naive controls in their stress response and all psychobiological systems have returned to their pre-stressor state. For example, Sakellaris and Vernikos-Danellis (1975) found that rats chronically exposed to cold have more rapid corticosteroid responses to other novel stimuli than do unexposed controls, even though they fail to show any adrenocortical responses to the original stressors after repeated re-exposure.

Summary

The studies of adaptation have implicated a neural suppression system involving amygdala nuclei. However, it is significant that there are marked individual differences in the original response and rates of adaptation which argues for differences in limbic function.

INDIVIDUAL DIFFERENCES

Early studies of the stress response emphasised the breadth

of neuroendocrine responses to a wide variety of stressors but neglected variation in the individual animal's response.

The following chronic avoidance study examined this question (Levine et al., 1972). Monkeys were given a 72 hour shock avoidance and they had not increase in adrenal cortical activation after the first few sessions; when chronically exposed they show even lower levels than baseline. There were marked individual differences in response, depending on social status. When dominant and aggressive rhesus monkeys were put in shock avoidance they showed a very rapid increase in adrenal cortical activity. However, submissive and less aggressive monkeys failed to show any increase of adrenal cortical activity when subjected to shock avoidance.

When the monkeys were separated, in terms of aggressiveness and speed of learning, the aggressive, slow to learn group showed evidence of continued emotionality (17-OHCS elevation) during the total experimental period, while the non-aggressive, quick to learn group showed little effect. Since the task was the same for both groups and their performance was similar, it was concluded that the variable producing this difference between the two goups was some neural mechanism that is related to aggressiveness. It is interesting that parts of the amygdala have been related to aggressiveness in a number of studies, e.g. Wiepkema et al. (1980).

A monkey's level of aggressiveness will be a function of both genetic factors and experience. No genetic evidence was available but Levine et al., speculated on the influence of experience. They argued that, for subordinate and submissive animals, avoidance of punishment is not as novel or as potentially stressful an event as it is for a dominant male.

On the issue of experience, they noted that the two monkeys with the highest 17-OHCS avoidance response in a previous study were "laboratory-naive" animals and the monkeys who showed the smallest reaction had extensive laboratory experience. It is possible that this prior history made it easier for some monkeys to adapt to prolonged exposure to the repeated avoidance. Clearly, an animal's past history is a relevant factor in the present response to a stressor.

CORTICAL CONTROL

Some stressors, such as pain or loud noises, require little neural analysis for eliciting the stress response. Pain probably has direct inputs to the hypothalamic regions, while a loud noise would directly activate the hippocampal formation. However, information about social status requires extensive analysis and this complex sort of experience must be mediated to a great extent by cortical systems. In the various sensory regions of the neocortex, sensory information is transformed by such factors as experience and context to give meaning. In this way the psychological meaning of the information for the individual is derived.

Complex events will be analysed and interpreted at the cortex first and any mismatch between input and expectancy identified and then the cortex will excite the hippocampus. Thus, simple expectancies only involve the hippocampus, but complex expectancies depend on cortical analyses (Warburton, 1979). There is a cascading of inputs from a number of cortical areas through all adjacent regions leading to the entorhinal cortex, suggesting that the hippocampus receives highly analyzed, abstracted information from all modalities, rather than information about any specific modality(Wilson, 1985).

There is also evidence for the involvement of the cortex with parts of the amygdala. The baso-lateral amygdala receives a plentiful input from the neocortex (either directly or indirectly), but this input comes mostly from the secondary sensory areas which, it is now known, deal with the processing or elaborations of information received from the primary ones (Ottersen, 1981).

INTEGRATED MODEL OF CONTROL

In this final section, the neural control systems will be integrated into a single model.

The changes in hormone excretion that are produced by stressors is controlled via neural pathways that converge on the hypothalamus from other parts of the brain. The control of corticosteroid secretion has been investigated most and will be considered first. It seems that there is a noradrenergic system inhibiting corticosteroid secretion, while a cholinergic pathway is initiating release. This pathway appears to be a branch of the cholinergic pathways controlling cortical arousal and the information processing changes which are part of the stress response.

The organism establishes simple and complex expectancies about the world around him, but novel information represents an increase in uncertainty and produces an increase in electrocortical arousal. The hippocampus is responsible for identifying low information stimuli and preventing electrocortical arousal feedback pathway to the midbrain region. The amount of the arousal will be a function of size of the mismatch between expectancy and input (see Pribram, 1967). This implies that there must be storage of information about simple, predictable stimuli in the hippocampal formation (Warburton, 1979).

Gray (1982) has suggested that the hippocampus is a comparator which processes information, which describes the current state of the world and information of expectancies about the future state from a generator of predictions. The generator of predictions uses information describing regularities in the past experience of the animal. These regularities are of two principal kinds, each mapping on to classical and instrumental conditioning: namely, stimulus-stimulus associations and response-stimulus associations. The comparator then decides whether there is a match or mismatch between the description of the world and the prediction.

If there is a mismatch between expected and actual events, the system stops behaviour and prepares for possible rapid and vigorous action. Co-ordinated with the change in information processing and motor intrruption, is the release of stress steroids. One output of the hippocampus is the medial corticohypothalamic tract from the presubiculum to the hypothalamus which could provide a direct pathway for hippocampal influence over the pituitary-adrenocortical system (Wilson, 1985) as well as the indirect input from the hippocampus to the hypothalamus by way of the septum to the origins of the cholinergic pathways (Warburton, 1979).

There is also evidence for the involvement of parts of the amygdala in control of the stess steroid response (Mason et al., 1961). It has been argued that the amygdala are concerned with the neural processes whereby motivationally-important decisions are taken, and so it must receive information about current sensory input (Herbert, 1984). Parts of the amygdala receive direct or indirect inputs from the secondary sensory areas of the neocortex where complex elaboration of information occurs. It is beginning to look as if the amygdala are an entrance to the limbic system for this kind of input, which can then act upon those other areas (e.g. septum, hypothalamus, and brainstem) which excite the relevant motivational response, in the light of the animal's estimate of 'costs' and 'benefits' (Herbert, 1984).

Finally, we come to the intriguing relation of the stress response with aggressiveness. Wiepkema et al. (1980) have argued that portions of the amygdala are responsible for determining whether a response will be suppressed or released on the basis of the individual's experience. If, as Herbert (1984) has argued, the amygdala are responsible for defining the costs and benefits, then it can be seen that the amygdala could be part of the systems by which an animal's social structure modulates behavioural responses.

ses. Subordinate males must take moment-to-moment decisions on the basis of their experinece about whether the 'benefit' of mating (in terms of immediate reward) is likely to outweigh the 'costs' of being attacked. This ability can be considered as a part of the 'coping' mechanism of subordination and this ability could be adaptive for the performance of shock avoidance.

REFERENCES

Endröczi, E., Hartemann, G. and Lissak, K., 1963a. Effect of intracerebrally administered cholinergic and adrenergic drugs on neocortical and archicortical electrial activity. Acta Physiol. Hung., 24, 207-209.

Endröczi, E., Schreiberg, G. and Lissak, K., 1963b. The role of central nervous activating and inhibitory structures in the control of pituitary-adrenocortical function. Effects of intracerebral and adrenergic stimulation. Acta Physiol. Hung., 24, 211-221.

Gray, J.A., 1982. "The Neuropsychology of Anxiety: An enquiry into the functions of the septo-hippocampal system". (Oxford University Press, Oxford).

Herbert, J., 1984. Behaviour and the limbic system with particular reference to sexual and aggressive interactions. In "Psychopharmacology of the Limbic System" (Eds. R. Trimble and E. Zarifian). (Oxford University Press, Oxford). pp. 51-67.

Krieger, H.P. and Krieger, D.T., 1970. Chemical stimulation of the brain: effect on adrenal corticoid release. Amer. J. Physiol., 218, 1632-1641.

Levine, S., Goldman, L. and Coover, G.D., 1972. Expectancy and the pituitary-adrenal system. In "Physiology, Emotion and Psychosomatic Illness" (Eds. R. Porter and J. Knight). (Elsevier, Amsterdam).

Mason, J.W., Nauta, W.J.H., Brady, J.V., Robinson, J.A. and Thach, J.S., 1960. Limbic system influences on the pituitary-adrenal cortical system. Psychosom. Med. 22, 322-336.

Mason, J.W., Nauta, W.J.H., Brady, J.V. and Taylor, E.D. , 1961. The role of the limbic system structures in the regulation of ACTH secretion. Acta Neuroveg. (Wien), 23, 4-10.

Mason, J.W., Brady, J.V. and Tolliver, G.A., 1968. Plasma and urinary 17-hydroxycorticosteroid responses to 72-Hr.avoidance sessions in the monkey. Psychosom. Med., 30, 608-630.

Mason, J.W., 1975. A historical view of the stress field. J. Human Stress, 1, 6-12.

Ottersen, O.P., 1981. The afferent connections of the amygdala of the rat as studied with retrograde transport of horseradish peroxidase. In "The Amygdaloid Complex" (Ed. Y. Ben-Ari). (Elsevier, New York(). pp. 91-104.

Pollard, I., Basset, J.R. and Cairncross, K.D., 1976. Plasma glucocorticoid elevation and ultrastructural changes in the adenohypophysis of the male rat following prolonged exposure to stress. Neuroendocrinol., 21, 312-330.

Pribram, K.H., 1967. The new neurology and biology of emotion: a structural approach. Amer. Psychol., 22, 830-838.

Sakellaris, P.C. and Vernikos-Danellis, J., 1975. Increased rate of response of the pituitary-adrenal system in rats adapted to chronic stress. Endocrinology, 97, 597-602.

Schiaffini, O., Motta, M., Piva, F. and Martini, L., 1971. Role of brain transmitters in the control of the pituitary-adrenal axis. In "Hormonal Steroids" (Eds. V.H.T. James and L. Martini). (Exerpta Medica, Amsterdam).

Selye, H., 1937. Studies on adaptation. Endocrinol., 21, 169-175.

Warburton, D.M., 1975. Modern biochemical concepts of anxiety. Int. Pharmacopsychiat., 9, 189-205.

Warburton, D.M., 1979. Physiological aspects of information processing and stress. In "Stress and Cognition" (Eds. V.H. Hamilton and D.M. Warburton). (Wiley, London). pp 33-66.

Warburton, D.M., 1981. Neurochemical bases of behaviour. Brit. Med. Bull., 37, 121-125.

Warburton, D.M., 1983. Towards a neurochemical theory of learning and memory. In "Physiological Correlates of Human Behaviour. Volume 1" (Eds. A Gale and J. Edwards). (Academic Press, London). pp 143-158.

Warburton, D.M. and Russell, R.W., 1969. Some behavioural effects of cholinergic stimulation in the hippocampus. Life Sci., 8, 617-628.

Warburton, D.M. and Wesnes, K., 1984. Drugs as research tools in psychology: cholinergic drugs and information processing. Neuropsychobiol., 11, 121-132.

Wiepkema, P.R., Koolhaas, J.M. and Olivier-Aardema, R., 1980. Adaptive capabilities of the nervous system. Progr. Brain Res., 53, 369-384.

Wilson, M.M., 1985. Hippocampal inhibition of the pituitary adrenocortical response to stress. In "Stress: psychological and physiological interactions" (Ed. S.R. Burchfield). (Hemisphere Publishing, Washington) pp. 163-184.

APPLICATIONS OF NEUROBIOLOGICAL STUDIES TO FARM ANIMAL WELFARE

D.M. Broom

Department of Pure and Applied Zoology
University of Reading
P.O. Box 228, Reading RG6 2AJ
Berkshire
U.K.

ABSTRACT

Welfare problems arise when individuals fail to cope with their environment or when coping is difficult. Both short-term handling and long-term housing and management effects are of importance in farm animals. Studies of farm animal behaviour show that these animals must have the same sorts of psychological constructs in the brain as those thought to exist in man and laboratory animals. Complex expectancies are used in sophisticated analysis of the environment and hence the same problems concerning uncertainty and frustration exist. Brain-behaviour links are sparsely studied in farm animals but work on catecholamine agonists and opiate-peptide receptor-blockers suggest interactions of stereotyped behaviour with dopaminergic pathways and analgesic opiate peptides. Work on responsiveness of confined sows indicates that a neural gating process resulting in lack of responsiveness may be involved in coping with adversity associated with such confinement.

INTRODUCTION

When considering the responses of farm animals, or any other animals, to their environment it is desirable to distinguish situations which have detrimental effects on the individual from those which do not. The ultimate criterion of what is detrimental is whether individual fitness is reduced but it is often possible to say only that a particular effect on an individual is likely to result in reduced fitness. The term "stress" is of little use if it merely implies "a deviation from the optimum of a particular parameter in response to a stimulus" (Block, 1985) so it is best to use it only when referring to the processes within an individual when environmental factors (stressors) lead to detrimental effects (Broom, 1983a; 1985). The welfare of an individual is its state with regard to its attempts to cope with its environment. An individual which does cope with the conditions in which it lives may do so very easily, using its various regulatory systems, or it may have to spend much time and energy in doing so. Hence welfare may be bad either if the individual fails to cope or if coping is difficult.

Environmental effects which are of importance in farm animal husbandry include both short-term (acute) handling and long-term housing and management effects. Farm animals may be driven from place to place, loaded into vehicles, transported in vehicles, physically abused, mixed with other individuals, subjected to veterinary procedures and exposed to environmental events such as sonic booms. Long-term conditions may involve confinement in a small space, absence of particular environmental stimuli, social isolation or confinement with aggressive conspecifics. The long-term conditions may include frequent or intermittent episodes such as extreme temperature, lack of food, lack of water, contact with predators, or any of the other events listed above.

The first questions about the relevance to farm animal husbandry of work on responses to complex behavioural situations (Warburton, 1986) are whether these animals encounter such situations and whether they are able to have the complex concepts described for man. Some situations are obviously adverse, for example severe injury occurs or the animal is deprived of water for a long period, but the effects of uncertainty and frustration on farm animals may also be severe. Further questions concern whether the neural pathways and transmitters are the same and whether there are the same links between problem situations and the behavioural or physiological symptoms which have been described for man and laboratory animals.

EXPECTANCIES, UNCERTAINTY AND FRUSTRATION

How are expectancies recognised? In some experimental studies, experimenters deduce which sensory inputs are of high or low probability to the subject and recognise an expectancy when there is no response to sensory inputs which are of high probability but there is an obvious response to an input of low probability. In other studies, characteristic behaviour is shown shortly before a predictable event but not at other times and hence expectancy is assumed to exist. There is evidence of both kinds of expectancies in farm animals. Pigs fed regularly at a certain time of day show easily recognisable behaviour during the hour before feeding. Cattle accustomed to obtaining access to a particular Callan-Broadbent feeding gate when they try to put their head into it show obvious behavioural responses if the gate does not work because it is jammed or because their transponder is missing. Cows which have experienced unpleasant veterinary treatment when entering a crush are often unwilling to enter that crush. This may occur despite no lack of willingness to enter on an earlier occasion and is presumably a consequence of an expectancy of further unpleasant treatment there. Studies of social behaviour often indicate that individual pigs, cattle or poultry recognise other individuals who have previously attacked or threatened them. Since this can occur in the absence of any current threat and may be quite different from initial naive approaches to that same individual, it can be assumed that an expectancy of future attack exists.

More complex expectancies may also exist, for example when mountain sheep graze a meadow and then return to it only after several weeks, by which time it has regrown (Favre, 1975). An expectancy on a different time scale was demonstrated in an experimental study of habituation by domestic chicks. Chicks which had ceased to respond to a light bulb which was illuminated for 10s every 30s showed an orientation reaction when the bulb was extinguished after 5s or when the illumination continued for 15s (Broom, 1968).

Uncertainty may occur after an expected input is not received or in a novel situation in which no precise expectancy exists. Situations where there is a possibility of a predator attack have a considerable effect on the behaviour of sheep. Measurements of heart-rate during the approaches of dogs show that heart-rate increases as the dog draws nearer (Baldock, 1985; Baldock and Sibly, in press). Similarly, hens show behavioural or heart-rate responses when people approach (Duncan and Filshie, 1979). Uncertainty about cues associated with feeding were found to affect pig behaviour in an experiment by Carlstead (1984). When a bell sometimes signalled the arrival of food and sometimes did not, the pigs showed more aggressive acts per unit time active than when food arrival was reliably signalled by a bell and a light. Many situations where action by a conspecific or by man may occur involve uncertainty and have effects on farm animal behaviour and physiology.

If the levels of most of the causal factors which promote a behaviour are high enough for the occurrence of the behaviour to be very likely, but because of the absence of a key stimulus or the presence of some physical or social barrier, the behaviour cannot occur, the individual may be said to be frustrated. Examples include a hen which is about to lay an egg but which has no materials with which to build a nest, or a hen which is prevented from reaching normally accessible food by a perspex cover. The responses to such frustration included pacing behaviour or excessive, prolonged sitting in the first instance (Wood-Gush, 1969; 1972; Brantas, 1980) and stereotyped pacing or attacks on conspecifics in the second instance (Duncan and Wood-Gush, 1971, 1972). The evidence for frustration implies that the animal has a concept of the action which it wishes to carry out.

All of the examples in this section provide evidence for the existence of the same sorts of psychological constructs in the brain as those thought to exist in man and laboratory animals.

NEURAL PATHWAYS AND TRANSMITTERS IN THE BRAIN

The cerebral cortex is particularly large and elaborately folded in cattle, sheep and pigs as compared with rodents or carnivores. The basic anatomy of the brain is similar but evidence of functioning pathways in the brains of farm animals is sparce. Whilst it seems likely that the hypothalamic-cortical-hippocampal-tegmental arousal-controlling loops are present in ungulates this is not certainly known. The brains of poultry and other birds show close parallels in their basic pathways with those of mammals even though there are anatomical differences. Much also remains to be discovered, or verified, about the roles of the various catecholamines and opiate peptides in the brains of farm animals. It is likely that the links between the hypothalamic-adrenal pathways and arousal are very similar in farm animals and laboratory animals. Responses to adversity are normally associated with high arousal, but the reverse is not necessarily true; so it would be wrong to equate high arousal with detrimental effects on the individual.

One brain-behaviour link which has been explored in some detail is that associated with stereotypies. Pigs or calves which are kept in close confinement, usually with food available for very little of the time, show abnormal, stereotyped behaviour. A stereotypy is a relatively invariate sequence of movements occurring so frequently in a particular context that it could not be considered to form part of one of the normal functional systems of the animal (Broom, 1983b). These activities have been linked with brain function in several ways. Firstly, Sharman and Stephens (1974) and Fry et al (1976) showed that subcutaneous or intravenous injection of the dopamine receptor agonist apomorphine led to an increase in the incidence of stereotyped sucking, licking and chewing movements in cows, calves, sheep and piglets. Conversely, piglets prevented from carrying out sucking and stereotyped snout rubbing have been found to have modified brain dopamine metabolism. In each of these studies there is some possible contribution of nutritional factors, but they show some correspondence with laboratory studies on the effects on stereotypies of another catecholamine agonist, amphetamine (Lyon and Robbins, 1975). Apomorphine also increased the incidence of stereotypies in pigeons (Deviche, 1985).

As Dantzer (1986) points out, the links between dopamine agonists and behaviour are not easy to interpret. There are complex interactions among the brain neurotransmitters and there are also links with brain opioid peptides. The pentapeptides metenkephalin and leuenkephalin and the much longer chain peptide beta endorphin are involved in analgesia but they also have many other functions. The enkephalins have an affinity for delta receptors in the brain whilst morphine and beta endorphin have an affinity for mu receptors. These mu receptors are blocked by the drug naloxone. Dry sows, which normally spent long periods engaged in stereotypies, reduced this behaviour but did not change their exploratory behaviour, when naloxone was injected subcutaneously (Cronin et al; 1985). Hence it is possible that stereotypies help in coping with the environment by regulation of arousal and modulation of motivational state (Forrester, 1980; Broom, 1981, 1983) via interactions with dopaminergic pathways and the action of analgesic opioid peptides.

BEHAVIOURAL AND PHYSIOLOGICAL ATTEMPTS TO COPE WITH ADVERSITY

Many of the examples of attempts to cope with adversity described by Warburton (1979; 1986) have parallels in farm animal studies. As described by Ladewig (1986) the adrenal medulla and cortex are active in a wide variety of situations. For example, pigs which are cold or which are chased show increased adrenal cortex activity (Baldwin and Stephens, 1973) and levels of adrenaline were higher in fattening bulls which were kept on a slatted floor or which were involved in active interactions (Unshelm et al, 1978). Adverse conditions may elicit several different kinds of coping attempt in an animal and there are often individual differences in which coping methods are used most. Examples may be taken from studies of dry sows which are confined by tethering or by keeping in small stalls since these results have parallels with studies of learned helplessness in laboratory animals. When sows are first confined they show vigorous activity in the first hour but later they show reduced activity and often develop stereotypies (Cronin et al, 1984). In both tethers and stalls, sows vary in the amount and type of stereotyped behaviour which they show (Dantzer and Mormède, 1981; Broom and Potter, 1984). Behaviour and adrenal activity may be alternative responses to adversity, for Dantzer and Mormède found that the tethered sows which played with a chain had lower cortisol levels than those which did not. Stereotyped behaviour is an active response by sows to confined housing conditions but passive responses also occur. The low levels of activity shown by many confined sows have been described by Ekesbo (1981). It has also been suggested that confined sows show "apathetic" behaviour, meaning that they are unresponsive to events in their environment (Wiepkema et al, 1983). In experiments to test the responsiveness of stall-housed sows (Broom in press, a,b) all sows were very responsive to food arrival at normal feeding time and to the provision of a few food pellets during the middle of the day when most sows were resting. Behavioural responses were also shown to the filling of a food hopper which would be emptied in the trough the following morning. Stimuli not related to food elicited little response, however. Sows videotaped when a stranger stood quietly in front of them showed little behavioural response and even when the stranger's face was only 25cm from the nose of the sow there was no significant effect on behaviour except more time spent looking at the observer.

In experiments in which sows which had been lying for 20 minutes, but which had their eyes open, had 200ml of water at air temperature dropped on their backs, there was a clear difference between stall-housed and group-housed animals (Table 1).

TABLE 1 Responsiveness of stall-housed and group-housed sows. Behaviour in 20 min. after stimulus presentation.

	Stall-housed	Group-housed	
Mdn. time sit or stand(s)	27.5	349	p=0.096
Mdn number of other activities	2.5	6.5	p=0.004
n	24	12	(2-tailed)

The sows were videotaped before and after stimulus presentation and would have been likely to continue lying for some time. Many stall-housed sows failed to sit or stand during the 20 minutes after water was dropped on them and most showed few other activities such as raise head, vocalise, body shake, sniff ground etc. The group-housed sows sat up or stood up after a median of 2s and showed many more activities, especially in the first 2 minutes. Hence stall-housed sows were less responsive than group-housed sows but there was much individual variation in response, for example 4/24 stall-housed sows sat or stood for more than 19 minutes and showed 6 or 7 other activities. Similarly, there was variation among group-housed sows for four of these sat or stood for less than one minute. Three of these animals had been housed in stalls before their previous farrowing, so the effects of experience in stalls is being investigated further.

A possible interpretation of these experiments on responsiveness is that some confined sows attempt to cope with their conditions by some neural gating process. This might reduce sensory input by an efferent threshold modifier, or reduce arousal, or attenuate response to input in some other way. The process does not operate where the input is closely associated with food. It could be that such a neural gating effect might operate via the action of analgesic peptides but this need not be so. Whatever the mechanism, lack of responsiveness is an indicator that the animal is having to modify its normal functioning considerably in order to cope with its environment so its welfare is bad.

A final conclusion based on a wide variety of studies of farm animals is that their brain and behaviour mechanisms are sophisticated and deserving of further study and much respect.

REFERENCES

Baldock, N.M. 1985. Heart rate and behaviour recorded in sheep during undisturbed conditions and various husbandry practices. (Ph.D. thesis, University of Reading).

Baldock, N.M. and Sibly, R.M. in press. Effects of management procedure on heart rate in sheep. Appl. Anim. Behav. Sci.

Baldwin, B.A. and Stephens, D.B. 1973. The effects of conditioned behaviour and environmental factors on plasma corticosteroid levels in pigs. Physiol. Behav., 10, 267-274.

Block, W. 1985. Survival on land. Biologist, 32, 133-138.

Brantas, G.C., 1980. The pre-laying behaviour of laying hens in cages with and without nests. In "The Laying Hen and its Environment" (Ed. R. Moss), Curr. Top. vet. Med. Anim. Sci., 8, 227-234. (Martinus Nijhoff, The Hague)

Broom, D.M. 1968. Specific habituation by chicks. Nature, Lond., 217, 880-881.

Broom, D.M. 1981. Biology of Behaviour. (Cambridge University Press, Cambridge).

Broom, D.M. 1983a. The stress concept and ways of assessing the effects of stress in farm animals. Appl. Anim. Ethol., 11, 79.

Broom, D.M. 1983b. Stereotypies as animal welfare indicators. In "Indicators Relevant to Farm Animal Welfare", (Ed. D. Smidt), Curr. Top. vet. Med. Anim. Sci., 23, 81-87. (Martinus Nijhoff, The Hague)

Broom, D.M. in press. Responsiveness of stall-housed sows. Appl. Anim. Behav. Sci.,

Broom, D.M. in press. Stereotypies and responsiveness as welfare indicators in stall-housed sows. Anim. Prod.

Broom, D.M. and Potter, M. 1984. Factors affecting the occurrence of stereotypies in stall-housed dry sows. In "Proc. Int. Cong. Appl. Ethol. Farm Animals". (Ed. J. Unshelm, G. van Putten and K. Zeeb) (K.T.B.L., Darmstadt). pp.229-231

Carlstead, M.K. 1984. The influence of predictability of feeding on the behaviour of grower pigs. In "Proc. Int. Cong. Appl. Ethol. Farm Animals" (Ed. J. Unshelm, G. van Putten and K. Zeeb). (K.T.B.L., Darmstadt). pp. 274-278.

Cronin, G.M., Wiepkema, P.R. and Hofstede, G.J. 1984. The development of stereotypies in tethered sows. In "Proc. Int. Cong. Appl. Ethol. Farm Anim." (Ed. J. Unshelm, G. van Putten and K. Zeeb). (K.T.B.L., Darmstadt). pp. 97-100.

Cronin, G.M., Wiepkema, P.R. and van Ree, J.M. 1985. Endogenous opioids are involved in stereotyped behaviour of tethered sows. Neuropeptides, 6, 527-530.

Dantzer, R. 1986 Behavioural, physiological and functional aspects of stereotyped behaviour: a review and a reinterpretation. J. Anim. Sci. 62, 1776-1786.

Dantzer, R. and Mormède, P. 1981. Pituitary adrenal consequences of adjunctive activities in pigs. Horm. Behav., 15, 386-395.

Deviche, P. 1985. Behavioural response to apomorphine and its interaction with opiates in domestic pigeons. Pharmacol. Biochem. Behav., 22, 209.

Duncan, I.J.H. and Filshie, J.H. 1979. The use of radio telemetry devices to measure temperature and heart rate in domestic fowl. In "A Handbook on Biotelemetry and Radio Tracking", (Ed. C.J. Amlaner and D.W. MacDonald) (Pergamon, Oxford). pp.579-588.

Duncan, I.J.H. and Wood-Gush, D.G.M. 1971. Frustration and aggression in the domestic fowl. Anim. Behav., 19, 500-504.

Duncan. I.J.H. and Wood-Gush, D.G.M. 1972. Thwarting of feeding behaviour in the domestic fowl. Anim. Behav., 20, 444-451.

Ekesbo, I. 1981. Some aspects of sow health and housing. In "The Welfare of Pigs", (Ed. W. Sybesma). Cur. Top. Vet. Med. Anim. Sci., 11, 250-264. (Martinus Nijhoff, The Hague).

Favre, J.Y. 1975. Comportement d'Ovins Gardes. (Ministere de l'Agriculture École Nationale Superieure Agronomique de Montpelier).
Forrester, R.C. 1980. Stereotypies and the behavioural regulation of motivational state. Appl. Anim. Ethol., 6, 386-387.
Fry, J.P., Sharman, D.P. and Stephens, D.B. 1976. The effect of apomorphine on oral behaviour in piglets. Br. J. Pharmacol., 56, 388p
Fry, J.P., Sharman, D.F. and Stephens, D.B. 1981. Cerebral dopamine, apomorphine and oral activity in the neonatal pig. J. Vet. Pharmacol. Therap., 41, 193.
Ladewig, J. 1986. this volume
Lyon, M. and Robbins, T., 1975. The action of central nervous system stimulant drugs: A general theory concerning amphetamine effects. In "Currents Developments in Psychopharmacology (Vol. 2)", (Ed. W. Essman and L. Valzelli). (Spectrum, New York). pp. 81-163.
Sharman, D.F. and Stephens, D.B. 1974. The effect of apomorphine on the behaviour of farm animals. J. Physiol., 242, 25P.
Sharman, D.F., Mann, S.P., Fry, J.P., Banns, H. and Stephens, D.B. 1982. Cerebral dopamine metabolism and stereotyped behaviour in early-weaned piglets. Neuroscience, 7, 1937.
Unshelm, J., Smidt, D., Andreae, U., Ellendorff, F. and Elsaesser, F. 1978. Haltungssysteme und soziale Rangordnung als Einflussfaktoren biochemischer Parameter. Tierhaltung, K.T.B.L., 233, 179-185.
Warburton, D.M. 1979. Physiological aspects of information processing and stress. In "Human Stress and Cognition" (Ed. V. Hamilton and D.M. Warburton). (Wiley, Chichester). pp.33-65.
Warburton, D.M. 1986. this volume.
Wiepkema, P.R., Broom, D.M., Duncan, I.J.H. and van Putten, G. 1983. "Abnormal Behaviours in Farm Animals." (Commission of the European Communities, Brussels).
Wood-Gush, D.G.M. 1969. Laying in battery cages. Wld. Poult. Sci. J., 25, 145.
Wood-Gush, D.G.M. 1972. Strain differences in response to sub-optimal stimuli in the fowl. Anim. Behav., 20, 72-76.

INTRODUCTION

Contrary to what could be expected, there is little systematic research available on the relationship between stress and its behavioural concomitants. In fact the mass of endocrinological data on stress is in no way in balance with a corresponding set of behavioural findings. Animal behaviour research either focussed on behaviour patterns performed under natural conditions with an emphasis on ontogeny, causation, function and evolution (the classic topics in ethology), or on learning characteristics and capabilities (the classical topics of comparative animal psychology). In both domains the concept of stress had only indirect significance.

Since stress refers to some individual state of being overloaded and therefore has an obvious relationship with the concept of animal welfare, and since the latter term is crucial in the area of applied ethology of farm animals, quite new connections should be made between the study of animal behaviour and modern stress research. It is self-evident that such an endeavour can only be successful in so far we succeed in integrating behavioural and physiological data and theories.

In his chapter Wiepkema takes a starting point in the datum that organisms have been adapted to their Umwelt in a very specific way. This specificity not only pertains to the way organisms react upon Umwelt changes. It also implies the development of Sollwerte (laid down in the brain), which enable them to actively regulate their environment according to these Sollwerte. As shown in behaviour research organisms are not entirely free in their means to regulate specific aspects of their Umwelt; this critical point has been analysed under the heading "Constraints of learning". Every time adaptation is problematic some form of stress will be present.

Besides the foregoing which is mainly derived from ethology, comparative psychology has informed us that learning basically represents the capability to associate causally related events. This opens the way to predict and/or to control environmental changes. Expectancies become plausible. Vertebrates appear to be goal directed systems having Sollwerte and expectancies (cf. also the contribution of Toates). Wiepkema proposes an integrated and operational description of stress in which the concepts of predictability and controllability are central. Behaviourally, stress appears to be expressed in conflict behaviours (as a result of acute stress) and in disturbed behaviours (as a result of chronic stress).

This line is taken up by Ödberg illustrating the actuality and significance of notably disturbed behaviours in the present day farm animal situation. Apart from the recurrent phenomenon of individual differences (see foregoing chapters), Ödberg also points to a largely overlooked finding and asks: Why do organisms prefer to do something when being stressed instead of "simply" stopping behaviour and waiting for better times?

Wiepkema makes a first attempt to incorporate the role of emotions in behavioural programs in order to obtain an integral model of vertebrate experiences of pleasure and displeasure (cf. Cabanac's chapter). His line of thinking closely parallels ideas put forward in a recent book of Simonov (1986). It is interesting to note that gradually the purely behaviouristic approach is replaced by one in which emotions of animals are no longer taboo. On the contrary, evidence is accumulating that emotions have great significance in shaping behaviour as long as the outcome of that behaviour is not entirely certain.

Applied ethology can only play a significantly explaining and practical role in farm animal welfare (and stress) matters, if it becomes integrated with recent developments in physiology (neuroendocrinology, immunology) and strongly furthers the development of cognitive ethology.

REFERENCES

Simonov, P.V., 1986. The emotional brain. Plenum Press, New York (translated from Russian).

BEHAVIOURAL ASPECTS OF STRESS

P.R. Wiepkema

Agricultural University,
Department of Animal Husbandry and Ethology
Marijkeweg 40, 6709 PG Wageningen,
The Netherlands.

ABSTRACT

In the course of evolution organisms have been adapted to a species-specific part of their environment. This part, their Umwelt, is continuously on the move; it is not only transformed during a lifetime, but it also changes from hour to hour and from day to day. All these shifts and changes are mainly non random. Vertebrates respond, influence and anticipate most of these Umwelt-changes by performing both fixed (routines), and flexible, behaviours. The organism is in a continuous interactional state with its dynamic Umwelt. Three processes underlying and shaping this interactional state are discussed: 1) ontogenetic processes (like imprinting) and their relevance for adult behaviour 2) learning or the capability to associate causally related events (conditioning and operant learning) and 3) constraints in adaptation (the need to perform species-specific behaviour programs). Most if not all of these behaviour programs are parts of regulatory systems that serve the control of relevant Umwelt measures. Such programs are like scenarios that include their own expectancies. During acute and chronic stress, expectancies are not fulfilled; there is a reduced predictability and/or controllability of relevant Umwelt changes. Stress results from uncertainties experienced during and inherent in a given interactional state. The amount of stress can be measured in terms of changes in transitional probabilities of successive events. During acute stress, conflict behaviours (e.g. redirected, ambivalent and displacement activities) arise, whereas chronic stress brings about disturbed behaviour (stereotypies, injurious activities etc.) The significance of individual differences in coping is discussed briefly. It is argued that all Vertebrates have homologous emotions that play a critical role in shaping behaviour programs. Long-lasting negative emotions occur during chronic stress and characterize disturbed welfare.

UMWELT

In the course of evolution, organisms have been adapted physiologically and behaviourally to those parts of our earth that favoured individual viability. Since these parts of the environment - ecological niches - have become species-specific, we must infer that individuals of a given species cannot survive in each arbitrary environment. There are constraints on adaptation. The natural environment of an organism is characterized by the fact that it consists of only elements that can be perceived by the organism or influenced through its behaviour. To indicate this specific environmental compartment, v. Uexküll (1956) coined the term "Umwelt". Full understanding of what happens when a given organism is overtaxed and, as a result of this, develops a state of stress, is only possible if we have adequate knowledge about the Umwelt in which that organism lives.

It is highly relevant that an Umwelt , embedding a given individual, is never a static entity, but in at least two respects is a very dynamic one. First, the individual's Umwelt changes drastically during its ontogeny: for instance the suckling's Umwelt is entirely different from the one that surrounds the same individual when being an adult mammal. Secondly, during each stage of life the Umwelt not only oscillates according to certain rhythms (e.g. diurnal, annual), but it also fluctuates as a result of the occurrence of all sorts of Umwelt events (the hatching of young, the arrival of a predator and so on). The important point is that, at the time base of the individual's life, all these fluctuations are mainly non-random; that is , most of these Umwelt events or Umwelt changes are announced somehow. Individuals have been adapted to these time characteristics of their Umwelt. The picture emerging from such considerations is that each healthy individual during each stage of its life neatly fits numerous dynamic aspects of its Umwelt in order to survive and to reproduce. This fitting ranges from the submicroscopic level up to and including the individual one.

Animals not only respond to relevant Umwelt events, they can also anticipate Umwelt changes; for instance, in active avoidance situations or when preparing themselves for migration (by being temporarily hyperphagic). Moreover, they may also influence and change their Umwelt; for instance, by constructing a well-insulated and camouflaged nesting site, or by driving an intruder out of their territory and so on. It is relevant to emphasize that, when interacting with a dynamic Umwelt, Vertebrates (presumably more than invertebrates) employ flexible behaviour programs in order to realise one and the same consummatory situation (Toates, 1986). This freedom of behaviour is characteristic for the (farm) animals we are discussing now. The same freedom might be reflected in the perplexing individual differences that can be observed when organisms cope with apparently the same sort of stressor.

Each time some Umwelt measure (for instance local temperature) shifts away from its ideal or optimal value the organism performs countermeasures that may restore the original situation (for instance moving to a place with a better temperature). The first time that the organism encounters a particular Umwelt shift, the shift will typically hold the power to evoke stress. The resulting stress-state (or stress response) is more serious the less the organism can restore the situation or the less it can adapt in a new way. I shall try to operationalize this type of stress, which is basically an interactional state of the organism and its Umwelt. Further, I shall indicate what sort of behaviour patterns are typical of stress. Finally, I want to comment on the question as to why Vertebrates may experience negative and positive emotions while coping.

A good understanding of the coping mechanisms involved is only possible if we first have a brief look at three particular and interwoven behaviour processes: ontogenetic processes, learning and the typical constraints of species-specific (instinctive) behaviour.

ONTOGENETIC PROCESSES

When a chick hatches or a young mammal is born they find themselves in the midst of a period in which the brain still forms new neural connections, while simultaneously the Umwelt of these animals expands enormously (cf. Bullock, 1977). In this period the organism receives, but also may more actively gather information that drastically and often irreversibly determines the final hardware of the brain; that is the final neural pattern of numbers and types of neurons and their interconnections. A striking example can be found in the work of Blakemore and others on the development of neurons in the visual cortex of cats (Blakemore, 1974).

Therefore, during early life, Umwelt events profoundly fasten down brain faculties operating during adulthood. In this way adult behaviour strongly depends on the information taken up in roughly the perinatal period. We can even go further and state that as a rule specific information is needed during specific sensitive or critical periods. Outside such a period the given information is non effective. Typical examples are the perinatal differentiating effects of testosterone on brain and adult behaviour of rats (Gorski, 1979;Goy, 1970) and the way certain birds learn their adult song (Kroodsma, 1982).

The information integrated during these early life stages mainly pertains to the development of adequate social skills like recognizing and interacting with conspecifics. Imprinting of non-conspecifies underlies domestication.

If this early and often crucial information is deficient, already in early life this may lead to the development of disturbed behaviour patterns (cf. Davenport, 1979) that probably signify serious forms of stress. Such abnormal behaviours have also been described in piglets (van Putten and Dammers, 1976; Schouten 1986) and in veal calves (van Putten and Elshof, 1982; Webster and Saville, 1982; de Wilt, 1985). I will return to this point later.

For the moment I want to focus on other and far reaching consequences of these early life deficiencies. Such deficiencies not only reduce quantity and quality of social interactions (cf. Hemsworth et al.,1977; Kruyt, 1964; Schouten, 1986), but they also reduce the coping faculties of the same organism. For instance, monkeys raised in an impoverished environment (artificial mother) are much less explorative and have much weaker learning capacities than conspecifics raised in a rich and normal environment (on a mother). Monkeys of the first group are scarcely attentive to what happens in their Umwelt (Mason, 1978) and in fact show typical signs of learned helplessness (Maier and Seligman, 1976). Informational shortcomings in early life may drastically and permanently diminish the organism's capacity to detect contingencies in its Umwelt. It is evident that such animals will have special and often very serious coping problems when later on they have to adapt to all sorts of fluctuating Umwelt variables. This holds especially when the actual environment strongly differs from the optimal one. Striking examples of this are found in modern husbandry systems.

LEARNING

As stated before, most if not all Umwelt variables fluctuate in a non-random way. This is the same as saying that most Umwelt changes do not occur unexpectedly. It makes sense to distinguish two types of Umwelt change: 1) those which are only predictable and 2) those which are also controllable.

1) Typical examples of the first category are changes in the weather, or the arrival of a predator. Signals often precede such changes and the brain is able to associate such successive events if there is a causal(non-random) relationship between them. This type of associative learning has been coined as conditioning.

2) The second type of Umwelt change depends on the perfomance of some specific activity. For instance, certain actions bring food within reach, others remove an unwanted intruder and so on. These relationships are based on learning processes in which an association is established between a behaviour pattern and its conjoint Umwelt changes: we call this operant learning . In brief, conditioning makes it possible for an organism to predict and to anticipate Umwelt changes, while in addition operant learning enables the same organism to control relevant Umwelt changes (cf. Toates, this volume).

There is good evidence that organisms not only associate successive events, but also assess the probability of successive events. (Dickinson, 1980). If we call the preceding event E_1 (being a signal or an operant) and the following event E_2 (being the arrival of a predator or the acquisition of food),then the probability that E_2 will follow E_1 can be described as $p.E_2/E_1$ (p. ranging from 0.0-1.0). Research on animal learning has also shown that not only can an organism assess $p.E_2/E_1$ but it can also assess the probability of an E_2 occurrence, when E_1 does not occur; that is $p.\ E_2/no\text{-}E_1$ (Dickinson, 1980). (Fig.1) A relevant Umwelt change (being E_1---->E_2)is then unpredictable and/or uncontrollable if and only if $p.E_2/E_1 = p.E_2/no\text{-}E_1$ (diagonal in fig. 1).

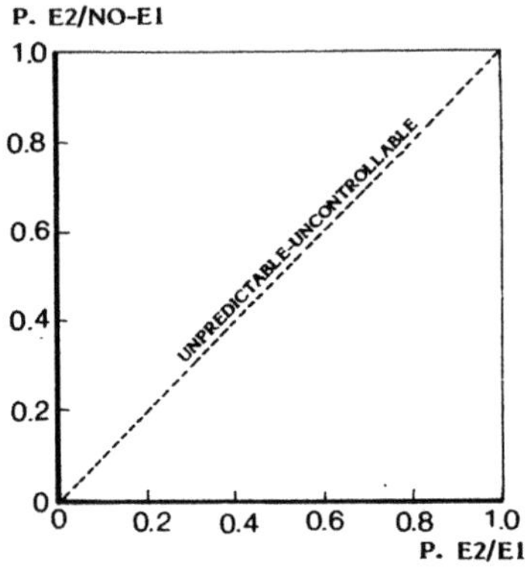

Fig. 1 Predictability - controllability space. $P.E_2/E_1$= conditional probability of event 2 following event 1. $P.E_2/no\text{-}E_1$= the conditional probability of event 2 following the non-occurrence of event 1.

Therefore, as a rule an experienced organism will have a rather specific and quantitative expectation of what shall happen following a given and observed signal or following the perfomance (or non-performance) of a given behaviour pattern. In other words, such organisms are able to predict and to control relevant Umwelt changes.

It is relevant to mention that an organism not only evaluates different E_2 states differently (as appears from choice experiments), but that it also can transpose and implement such knowledge from one situation to another. For example, rats that in one situation have learned that a specific food is hazardous, are able to transpose and to implement this knowledge directly into another situation where the same food is also present (Dickinson , 1985).

A final remark with regard to learning and associated behaviour programs concerns the point that, if an observed signal predicts, or a certain operant, controls a given Umwelt change over and again with high predictability ($p.E_2/E_1 = 1.0$), the behaviour patterns involved will become routines or habits; actions change into habits (Dickinson, 1985). The older a habit the less it depends on Umwelt fluctuations, the less flexible its performance and the more difficult to reshape it when needed (as shown during extinction procedures). These aspects characterize the observable differences between actions and habits. Later on I shall argue that both types of behaviour may also have quite different emotional loadings.

CONSTRAINTS OF BEHAVIOUR

In many cases, it is not only important for the organism to reach a goal situation (for instance being in touch with adequate food), but also the means by which this situation has been attained matters (cf. Toates, this volume). For instance, when given the choice between free food and food for which the animal has to work (bar pressing) rats prefer the latter condition (Singh, 1970). Comparable data have been described for chickens (Duncan and Hughes, 1972) Obviously some introductory and active program should precede the acquisition of food (and presumably prepare the organism for an optimal processing of this food).

We can even go further and note that in, say, pigeons, the response of key pecking for food contains feed intake components, whereas key pecking for water contains water intake components (Moore,1973). This type of phenomenon has been described under the heading "autoshaping". For some unknown reasons the attainment of specific goal situations implies specific preceding programs. This statement is closely related with the finding that an organism cannot associate any arbitrary signal (or activity) - being E_1- with any event (E_2) following thereupon. There are species-specific constraints in associating or learning (cf. Shettleworth, 1975).

If this reasoning is true one could expect that , if

organisms cannot or need not perform essential parts of their behavioural program to attain some specific goal (for instance food), such organisms will compensate for this by performing the missing elements in an out-of-context, and therefore abnormal, way. This is exactly what happens in training situations (Breland and Breland, 1961), in Zoo animals (Morris, 1964) and present-day farm animals (e.g. Mees and Metz, 1984; Ruiterkamp, 1985; Schouten, 1986; de Wilt, 1985).

A crucial question is then why organisms do perform "superfluous" behaviour and do not simply omit or suppress such behaviour patterns. Until now, we have no satisfactory explanation.

PREDICTABILITY AND CONTROLLABILITY

Much of the foregoing is summarized by stating that an organism can be represented as an information processing system; however, this system shows species-specific restrictions. The information itself serves the maintenance of the individuals, which all have their own and often very specific demands upon their Umwelt.

In order to make my line of thought as clear as possible I have to distinguish at least two types of information: 1) Sollwert and 2) expectancy.

Sollwert.

This first category includes all the neural information that indicates which value a given Umwelt measure should have; for instance the optimal value of body temperature or the detailed structure of a nest for laying and brooding eggs. I call this set of information Sollwerte (= plural of Sollwert) or standards; Sollwerte have been formed in the course of evolution and are therefore strongly similar for individuals of the same species, sex and age. Each Sollwert corresponds with an Iswert, being the actual Umwelt value of that Sollwert. The comparison of a perceived Istwert and its corresponding neural Sollwert is the core of the information processing involved here. If there is a mismatch between Istwert and Sollwert, a program that reduces this difference will be activated and may be performed: a negative feedback system that regulates essential Umwelt values (cf. Wiepkema, 1985). The difference between Istwert and Sollwert is called the motivation of the program performed.

To prevent misunderstandings, I have to add that Sollwerte do not refer to one and only one value of a given Umwelt measure. A Sollwert probably refers to a range of acceptable values as determined by and depending on endogenous rhythms, actual Umwelt information and the strength of competing motivations.

Again this first category of information indicates which value a given Umwelt measure should have and, in fact, by this it defines the actual needs of an organism.

Expectancy

The second category of neural information contrasts with the first one in at least two aspects. First it refers to what an experienced animal expects to happen when a given signal is observed or a certain operant is performed (the E_1---->E_2 sequence). Secondly, the expectancies about E_2 may differ from individual to individual (even for the same species, sex and age) and can be reformed drastically on the basis of learning processes (cf. the learning paragraphs before).

Somewhat simplified, one could say that Sollwerte represent the irreversible information held by the species, whereas expectancies represent the reversible information held by the individuals of that same species. Both types of information enable the organism to control Istwert-Sollwert differences and to predict/control risky or profitable Umwelt events.

At this point I want to introduce the two key concepts of modern stress research: predictability and controllability of Umwelt changes. A reduced predictability or controllability appears to be the main cause of typical stress symptoms like elevated corticosteroid levels, stomach damage, conflict behaviours, utterance of pain etc. (cf. Levine and Ursin, 1980)

Stress refers to that interactional state (between organism

and its Umwelt) that is characterized by some form of uncertainty resulting from existing stressors (cf. Toates, this volume). Behaviourally this state can be operationalized as follows. Acute stress occurs if, and only if, predictability and/or controllability of relevant Umwelt changes decrease rapidly and significantly. Chronic stress takes place if predictability and/or controllability of relevant Umwelt changes remain very low during a long period. The shifts in predictability and/or controllability can be measured as changes in the transitional probabilities of successive events (E_1---->E_2). Such changes and their effects on behaviour have often been studied in learning experiments when modifying a reinforcement schedule; for instance when changing from a fixed ratio to a variable ratio schedule or during extinction procedures.

The intensity of the occurring stress is not only dependent on the jump in probability and/or the final low value of predictability/controlability, but also on the biological significance of the Umwelt change involved. This latter significance - be it positive or negative - has to be measured independently; for instance, in choice experiments or by recording how much work the animal is willing to perform in order to realize or to prevent that Umwelt change.

Under natural conditions, when the organism has its complete behavioural repertoire at its disposal, acute stress is the more common form of stress. If the worst comes to the worst the organism may simply flee and leave the disturbing situation. Chronic stress will occur when the organism is severly restricted in its behavioural possibilities; for instance when being confined.

Although in the following I shall focus on what happens when predictability and/or controllability decrease for relevant Umwelt changes - evoking stress - it is equally important to consider what may happen when the reverse is the case; that is , when predictability and/or controllability increase suddenly for relevant Umwelt changes. I shall not discuss which behavioural and physiological measures may signify a reduction of stress.

CONFLICT AND DISTURBED BEHAVIOURS

In the ethological literature (Hinde, 1970; Brown, 1975) the concept of stress is either absent or when present used in the classical physiological sense as introduced by Selye (Barnett, 1981; Huntingford, 1984). However this same literature offers excellent descriptions of a range of conflict behaviours (Hinde, 1970). These behaviours can easily be coordinated with (acute) stress, if we take uncertainty of the organism as a basic characteristic of stress (as argued just before). It is out of place to review conflict behaviour and their associated situations, as described in ethology, exhaustively. Some representative examples must suffice.

First there is the often described approach-avoidance conflict occurring for instance in males that defend their territory against some intruder. Inside the territory, attack dominates, whereas outside it, flight prevails. Just on the border where both tendencies are in balance, conflict behaviours occur. Secondly I mention the "lack of success" situation (or frustration), in which a given, and until then reliable, operant does not bring about its expected and normal consequence, e.g. obtaining food. In both situations, future events are uncertain and comparable conflict behaviours become overt. Agonistic behaviour may be performed by which the organism may free itself from the conflict situation. Another option is the performance of redirected - and/or the so called displacement - activities. There is good evidence that these latter types of behaviour may also be significant with respect to coping with the conflict situation itself. First they may have some signal value for conspecifics also involved in the conflict (Baerends, 1975; Hinde, 1970). Secondly they may result from autonomic processes activated during the conflict and simply represent for instance themoregulatory activities (Andrew, 1956). Further they may calm down the organism (Dantzer and Mormede, 1983) and/or shift its attention away from the conflict (McFarland, 1966).

An often overlooked point is that, in many conflict

situations, organisms may perform conflict behaviours which are specific for the conflict involved. A case in point are fighting cocks as described by Kruyt (1964); why should the future winner interrupt its agonistic behaviour for ground pecking, whereas the future loser interrupts with preening? The nature of the causal relationship between conflict state (stress) and its specific behaviour concomitants is largely unexplained.

A third example of a conflict and resulting behaviour is found in the work of Pavlov (1927), where he describes examples of acute neurosis in dogs. When discrimination learning became too difficult some dogs collapsed and started to squeal and tried to escape. Learning anew of the same task appeared to be very, very difficult for these dogs. Obviously, when the situation becomes too complex or too insecure conflict behaviour may occur; some of these behaviours may be interpreted as signs of fear and anger. I want to add a final category of conflict and its associated behaviour utterances, namely that of wounded or injured organisms. Such organisms may, for example, scream or perform ear and tail postures that probably express fear and uncertainty. There is little systematic information about when, how and why these behaviour elements are performed when being wounded. Nevertheless they may be highly significant during stress.

Similarly little information exists on what happens behaviourally when a conflict is unsolvable and therefore may become chronic. All the present evidence indicates that under such conditions organisms start to repeat and to ritualize their original conflict behaviours (Cronin, 1985; Morris, 1964; Stolba et. al. 1983). The final result may be the endless performance of stereotypies known from zoo animals and farm animals in modern intensive husbandry systems. However, under these chronic stress conditions not only stereotypies develop, but also injurious behaviour patterns by which organisms may mutilate themselves or conspecifics (Goosen and Ribbens, 1980; Wiepkema et al., 1983). In these cases a crucial point seems to be the impossibility of performing species-specific programs in order to reach a relevant goal situation (often the intake of food) (cf. Blokhuis and Arkes, 1984; Ruiterkamp, 1985; de Wilt, 1985).

An intriguing point is that, for a given animal a given chronic conflict situation may evoke more than one disturbed behaviour; each of these disturbed behaviours probably reflects different aspects of the same total conflict situation; for instance spot-picking and route-tracing of caged canaries reflect different aspects of the confinement canaries may live in (Keiper, 1969). Another example of this is found in veal calves kept in boxes and fed with milk only (Wiepkema et al. in prep.) Such calves develop different stereotypies each with their own time course. One stereotypy, biting/licking, of these calves has its optimum during the first 10 weeks of life, whereas a second stereotypy, tongue-playing starts to increase in the second 10 weeks of life. Moreover, the frequency of tongue-playing has a significant negative relationship with the frequency of abomasal (stomach) ulcers in the same animal, whereas such a relationship does not exists between biting/licking and abomasal damage. A plausible hypothesis is that different stereotypies reflect different aspects of stress existing more or less simultaneously in the same animal. What these specific relationships mean in terms of causation cannot be answered at this moment.

INDIVIDUAL DIFFERENCES

Great significance can be attached to the finding that individuals of the same species, sex and age may differ strongly in the way they cope with apparently the same complex of stressors (Corson and Corson, 1976). For rats Koolhaas and colleagues (1983) showed that during social conflict situations some adult males can be described as active copers, others as passive ones. The more general meaning of this variation is still unclear. However, such differences are highly relevant when evaluating chronic stress in individual animals. The following example will illustrate this point. As stated before veal calves may develop tongue-playing and may show abomasal ulcers at the day of slaughtering (age 20-22 weeks). The interesting point is that those animals which do not develop tongue-playing have one or more ulcers (or scars; the remnants of an ulcer) in the pyloric region of the abomasum. Those calves that do develop tongue-playing have absolutely no ulcers or scars at the same age (fig.2).

If we assume that ulcers are also unwished and have a negative meaning with respect to animal welfare, then the question arises as to which calves are in the worst state - those with stereotypies or those with ulcers? Again, as long as we do not understand why different individuals may follow different alternatives to cope with apparently the same stressor, we cannot formulate an adequate answer.

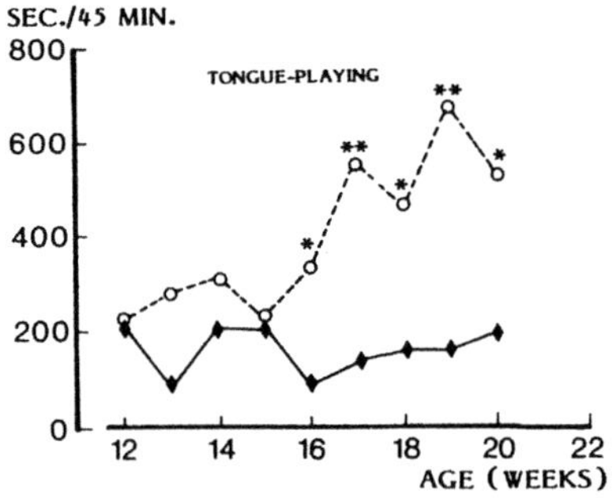

Fig.2 Changes in the median performance of tongue playing (ordinate) during the age of 12 - 20 weeks (abcis). o-----o veal calves with intact abomasum or erosions only (n=14); ♦——♦ veal calves with ulcer or scar (n=13). Both groups of calves differ significantly from week 16 on: * $P < 0.05$; ** $P < 0.01$.

EMOTIONS AND ANIMAL WELFARE

The rather naive assumption that men and other Vertebrates have comparable sensory and experiential capabilities has helped very much in designing experiments that have elucidated for instance how animals may identify and defend their territory, or how conspecifics may recognize and interact with each other, or why it pays to camouflage yourself or to mimic other organisms. Obviously Vertebrates have comparable sensory perceptions and sensations, which presumably result from profound homologies of structure (for instance in the brain) and of functions (for instance in behaviour systems) of these same Vertebrates.

The idea of homologies in behaviour is an old one and finds its roots in the comparative work of Whitman and Heinroth (cited by Lorenz, 1950) in the beginning of this century. Later on Lorenz (1950), Tinbergen(1951) and Baerends(1958) all emphasized that these homologies not only concern the morphology of behaviour patterns, but also their underlying mechanisms. In modern terms we should say that most, if not all, brain - behaviour interactions, or in other words, most, if not all, information processing events are strongly homologous in Vertebrates.

These homologies make it very likely that Vertebrates will also have homologous sensations and experiences each coupled with specific perceptive systems or combinations thereof; examples are visual and auditory ones, touch and pain. It is promising to class emotions or feelings among these experiences.

Emotions were not in vogue in classic ethology: it was held that they could not play a role in the causal explanation of behaviour (Tinbergen, 1951). This behaviouristic approach promoted the neglect of behaviour patterns that might express emotions; it is significant that ethologists paid relatively little attention to activities like, for instance, tail wagging or whining that could express pleasure or displeasure of the animals involved.

This attitude had much to do with the idea that nothing can be said about subjective experiences in animals or other individuals. Do two individuals, when responding to the same red colour, also undergo the same "subjective experience" of red? In my opinion this question is entirely irrelevant as long as we cannot operationalize the concept of "subjective experience". As long as this remains true, the concept is also irrelevant with respect to the assumption that individuals of the same and related species experience homologous (comparable) emotions or feelings. A more fruitful question is then, why an organism should have feelings of pleasure or displeasure (cf. Cabanac, this volume).

My colleague Frijda -a psychologist- and I have tried to work out the hypothesis that emotions - be they positive or negative - are closely related with strengthening or correcting behaviour programs. Moreover, they should have much to do with the key concepts of stress; predictability/controllability (Frijda, 1984; Wiepkema, 1982; 1985) The core of our reasoning is that individual organisms (Vertebrates) not only have Sollwerte (eigen-interests), but also alternative programs to realize one and the same Sollwert. Moreover, individual Vertebrates have expectations about the outcome of the successive steps in the programs used.

The role of emotions in normal behaviour can be described then in the following way (fig.3).

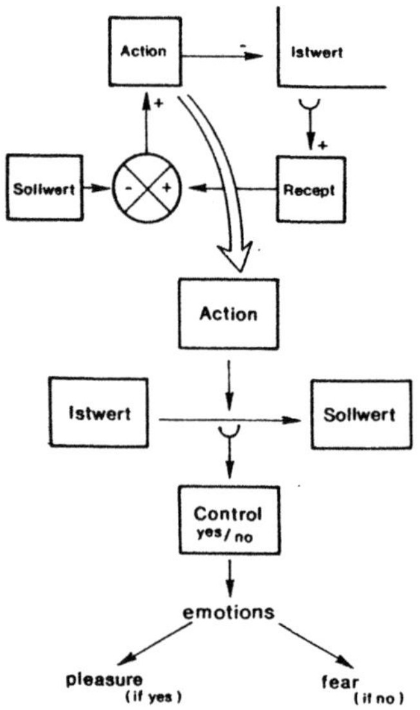

Fig.3 Role of emotion. The upper part of the figure represents the regulatory model described in the text. If an action (program) performed does indeed shift Istwert to its corresponding Sollwert, positive emotions arise. If not, negative emotions originate.

If an organism perceives a difference between an Istwert and its corresponding Sollwert, it will be motivated to reduce this difference (especially when it concerns a relevant Umwelt measure). This motivation sets into action a physiological and ethological program that in previous but comparable situations appeared to be an adequate instrument. However, the organism cannot be safe on this latter point, since the Istwert-Sollwert difference always takes place in a more or less fluctuating Umwelt. Events seldom repeat themselves exactly. Therefore the organism needs a monitoring system that adequately controls whether or not the program used (and its separate steps) really reduces the difference. This type of control system has been described also on the basis of neurobiological evidence (cf. Cools, 1985). Formally such a monitoring system can be described as a perceptive one. Behaviour programs are then much like scenarios inclusive their sequential expectancies.

The sensations coupled with monitoring the effect of a program (and its successive steps) may be called the organism's emotions. Positive emotions arise when the effect is the expected one or even better than that; such emotions strengthen the program used and promote its reuse in a future and comparable context. Negative emotions may stop and correct the program used and by this may facilitate the development of a new, alternative program.

The intensity of these emotions will be stronger 1) the greater the difference between expectation and real Umwelt change (E_1---> E_2) and 2) the higher the biological significance of the Umwelt change itself. Moreover, it is plausible that the further ahead an organism is able to foresee future events, the earlier his emotions may anticipate such events.

If the foregoing reasoning is true, emotions should play a dominant role only in those (parts of) behaviour programs of which the outcome is not entirely fixed. Such a program can be known by the relatively low predictability of its successive steps or, in other words, by its relatively high flexibility. When the same situation presents itself over and again flexible programs may change into routines, or actions into habits (cf. Dickinson, 1985). The conclusion might be then that routines and habits, because of their highly predictable/controllable outcomes, are no longer associated with emotions; organisms may perform such type of behaviour programs practically emotionless.

This reasoning also implies that, on the average, older organisms behave less emotionally than younger ones. Further, in Vertebrates emotions may be more common than in invertebrates, in so far the former behave less automaton-like than the latter.

CONCLUSION

The foregoing can be summarized by stating that acute stress implies an acute expectational deficit and by this an acute but temporary emotional experience. Chronic stress, however, brings about a chronic expectational deficit. This leads to long-lasting negative emotions and therefore to disturbed welfare. Such states of the organism are frequently found in modern farms and zoos, where animals find themselves in conditions of a chronic loss of controllability.

REFERENCES

Andrew, R.J., 1956. Some remarks on behaviour in conflict situations, with special reference to Emberiza spp. Brit. J.Anim. Behav. 4, 41-45.

Baerends, G.P., 1958. Comparative methods and the concept of homology in the study of behaviour. Arch. Neerl. Zool. 13, 401-417.

Baerends, G.P. 1975, An evaluation of the conflict hypothesis as explanatory principle for the evolution of displays. In:"Function and evolution in behaviour" (Eds. G.P. Baerends, C. Beer and A. Manning). Clarendon Press, Oxford, pp 187-227.

Barnett, S.A. 1981. Modern ethology. Oxford University Press, Oxford.

Blakemore, C., 1974. Developmental factors in the formation of feature extracting neurons. In :"The Neurosciences. Third study Program" (Eds. F.O. Schmidt and F.G. Worden). MIT Press, Cambridge, USA. pp 105-113.

Blokhuis, H.J. and J.G. Arkes, 1984, Some observations on the development of feather-pecking in poultry. Appl. Anim. Behav. Sci. 12, 145-157.

Breland, K. and M. Breland, 1961, The misbehavior of organisms. Am. Psychologist. 16, 681- 684.

Brown, J.L., 1975, The evolution of behaviour. Norton, New York.

Bullock, T.H. 1977, Introduction to nervous systems. Freeman, San Francisco.
Cools, A.R., 1985. Brain and behaviour: hierarchy of feedback systems and control of its input. In: Perspectives in ethology, vol 6 (Eds. P.H. Klopfer and P. Bateson), Plenum Press, New York, pp 109-168.
Corson, S.A. and E.O.L. Corson, 1976. Constitutional differences in physiological adaptation to stress and distress. In: G. Serban (Ed.) Psychopathology of human adaptation. Plenum Press, New York, pp 77- 94.
Cronin, G.M., 1985, The development and significance of abnormal stereotyped behaviours in tethered sows. Ph.D Thesis, Agricultural University, Wageningen.
Dantzer, R. and P. Mormede, 1983, De -arousal properties of stereotyped behaviour: evidence from pituitary- adrenal correlates in pigs. Appl. anim. Ethol. 10, 233-244.
Davenport, R.K. 1979, Some behavioural disturbances of great apes in captivity In:" The great apes" (Eds. A. Hamburg and E.R. McCown). Benjamin/ Cunnings, Menlo Park, pp 341-357.
Dickinson, A. 1980, Contemporary animal learning theory. Cambridge University Press, Cambridge.
Dickinson, A. 1985, Actions and habits: The development of behavioural autonomy. In:" Animal intelligence" (Ed. L. Weiskrantz). Clarendon Press, Oxford, pp 67-78.
Duncan, I.J.H. and B.O. Hughes, 1972, Free and operant feeding in domestic fowls. Anim. Behav. 20, 775-777.
Frijda, N.H., 1984, Towards a model of emotion. In:"Stress and anxiety" (Eds, C.D. Spielberger, I.G. Sarason and P. Defares), vol.9. pp 3-16.
Goosen, C. and L.G, Ribbens, 1980, Autoaggression and tactile communication in pairs of adult stumptailed macaques. Behaviour, 73, 155-174.
Gorski, R.A., 1979, Long-term hormonal modulation of neuronal structure and function. In:" The Neurosciences, Fourth study Program". (Eds. F.O. Schmidt and F.G. Worden) MIT Press, Cambridge, USA, pp 969-982.
Goy, R.W. 1970, Early hormonal influences on the development of sexual and sex-related behaviour. In : The Neurosciences, Second Study Program. (Ed. F.O. Schmidt) MIT Press, Cambridge, USA, pp 196-207.
Hinde, R.A. 1970, Animal behaviour McGraw-Hill, New York.
Huntingford, F. 1984, The study of animal behaviour. Chapman and Hall, London.
Keiper, R. 1969, Causal factors of stereotypies in caged birds. Anim. Behav. 17, 114-119.
Koolhaas, J.M., T. Schuurman and D.S. Fokkema, 1985. Social behaviour of rats as a model for the psychophysiology of hypertension. In: T.M. Dembrovski, T.H. Schmidt and G. Blümchen (eds.) Biobehavioural bases of coronary heart disease. Karger, Basel, pp 391-400.
Kroodsma, D.E. 1981, Ontogeny of birdsong. In:" Behavioural Development (eds. K. Immelman, G.W. Barlow, L. Petrinovich and M. Main). Cambridge University Press, pp 518- 532.
Kruyt, J. P., 1964, Ontogeny of social behaviour in Burmese Red Junglefowl (Gallus gallus spadiceus). suppl XII, Behaviour, 1-201.

Levine, S. and H. Ursin. (Eds.), 1980. Coping and Health, Plenum Press, New York.
Lorenz, K.Z., 1950. The comparative method in studying innate behaviour patterns, Symp. Soc. Exp. Biol. Cambridge, 4, 221-268.
Maier, S.F. and M.E.P. Seligman, 1976. Learned helplessness: theory and evidence. J. Exp. Psychol. General. 105, 3-46.
McFarland, D.J., 1966. On the causal and functional significance of displacement activities. Zschr. f. Tierpsychologie, 23, 217-235.
Mees, A.M.F. und J.H.M. Metz, 1984. Saugverhalten von Kalbern. Bedurfnis und Befriedigung bei verschiedenen Trankesystemen. KTBL-Schrift, 299, 82-91.
Moore, B.R. 1973. The role of directed Pavlovian reactions in simple instrumental learning in the pigeon . In:"Constraints on learning" (Eds. R.A. Hinde and F. Stevenson-Hinde), Academic Press, London, pp 159-188.
Morris, D. 1964. The response of animals to a restricted environment. Symp. Zool. Soc. London 13, 99-118.
Pavlov, I.P., 1927. Conditioned reflexes. Dover, New York.
Van Putten, G. and J. Dammers, 1976. A comparative study of the well-being of piglets reared conventiomally and in cages. Appl. Anim. Ethol. 2, 339-356.
Van Putten, G. and W.J. Elshof, 1982. Inharmonious behaviour of veal calves. In:"Disturbed behaviour in farm animals". (ed. W. Bessei). Hohenheimer Arbeiten, 121 pp 61-71.
Ruiterkamp, W.A. 1985. Het gedrag van mestvarkens in relatie tot huisvesting. Ph.D. Thesis, State University Utrecht, (English summary).
Schouten, W.G.P., 1986. Rearing conditions and behaviour in pigs. Ph.D. Thesis, Agricultural University, Wageningen.
Shettleworth, S.J., 1975. Reinforcement and the organization of behaviour in golden hamsters: hunger, environment and food reinforcement. J. Exp. Psychol: Animal Behaviour Processes. 1, 56-87.
Singh, D.,1970. Preference for bar pressing to obtain reward over freeloading in rats and children. J. comp. physiol. psychol. 73, 320-327.
Stolba, A., N. Baker and D.G.M. Wood-Gush, 1983. The characterization of stereotyped behaviour in stalled sows by information redundancy. Behaviour, 87, 157-182.
Tinbergen, N., 1951. The study of instinct. Clarendon Press, Oxford.
Uexkull, J. von, 1956. Streifzüge durch die Umwelten von Tieren und Menschen, Rohwolt, Hamburg.
Toates, F.M., 1986. Motivational systems. Cambridge University Press
Webster, A.J.F. and C. Saville, 1982. The effect of rearing systems on the development of behaviour in calves. In:"Welfare and husbandry of calves". (ed. J.P. Signoret). Nijhoff, Den Haag, pp 168-184.
Wiepkema P.R., 1982. On the identity and significance of disturbed behaviour in Vertebrates. In:"Disturbed behaviour in farm animals" (ed. W. Bessei). Hohenheimer Arbeiten, 121, pp 7-17.

Wiepkema, P.R., 1985. Abnormal behaviour in farm animals: ethological implications. Neth. J. Zool 35, 279-299.

Wiepkema, P.R., D.M. Brown, I.J.H. Duncan and G. van Putten, 1983. Abnormal behaviour in farm animals, CEC.-report, pp 1-16. Brussel.

Wiepkema, P.R., K.K. van Hellemond, P. Roessingh and H. Romberg. Behaviour and abomasal damage in individual veal calves. Submitted for publication.

Wilt, J.G. de, 1985. Behaviour and welfare of veal calves in relation to husbandry systems. Ph.D. Thesis, Agricultural University, Wageningen.

BEHAVIOURAL RESPONSES TO STRESS IN FARM ANIMALS

F.O. Ödberg

State University of Ghent
Faculty of Veterinary Medicine
Laboratory of Animal Genetics and Livestock Production
Heidestraat 19, B-9220 Merelbeke (Belgium)

ABSTRACT

If applied ethology has other aims than fundamental behavioural sciences and deserves acknowledgement on its own, it should also strengthen its contact with them, as it has worked too much in a theoretical vacuum with sometimes a deficient methodology; hence maybe its slowness in giving concrete advise to zootechnicians and lawyers.

Redirection activities, stereotypies and apathy are analysed as responses to chronic irreducible mismatches between Istwert and Sollwert. If such abnormal behaviours do have a homeostatic function, it would be advisable that future research should, 1º study developmental processes, 2º integrate closely behavioural and neurophysiological parameters, 3º take into account adaptation at different levels, and 4º focus on individual differences.

Does "coping" warrant welfare? One is confronted with the paradox, that neither the absence of abnormal behaviour implies no stress, nor their presence indicates actual stress. The argument that abnormal behaviours represent coping mechanisms should not induce us to take too quickly welfare for granted as, 1º it is likely that they only represent the final stage of a distressing process, 2º one still knows little about the costs of coping, 3º not all individuals in the same environment are able to develop such mechanisms.

INTRODUCTION

Some authors have published excellent reviews on the stress concept itself and on its relation with behaviour in farm animals (e.g. Dantzer and Mormède, 1983a; Friend, 1980) and it would be redundant to write a new version here. However, it is striking that most authors dealing with stress and behaviour do not push the analysis further once the following question arises (I keep using the cybernetical vocabulary used by Wiepkema in the preceding paper): "What happens when, after the detection of a mismatch between Istwert and Sollwert, a series of alternative responses have been tried out without success?". Gray (1982, p. 282) sugests that if the mismatch is of no fundamental consequence to the organism, the comparison stops and attention is geared to other requirements. If the mismatch is important and "none of them (alternative response routines) is successful, exploratory diversification of behaviour none the less eventually comes to an end, the animal typically engaging in some form of unlearned behaviour (e.g. grooming, sleeping)".

It is quite typical that this sentence represents only 5 lines in a book of 462 pages containing an extremely dense and rich analysis of the septo-hippocampal function. This author only hints at displacement activities and does not even mention any well-known abnormal behaviour. Even in the fundamental behavioural sciences there is indeed a remarkable lack of investigations on the mechanisms and functions of "abnormal behaviours"; unfortunately, these are just the questions the students of animal welfare are being asked.

McBride and Craig (1985) have recently argued that applied ethological research too often lacks a theoretical base. Wiepkema's model emphasizes the concepts of predictability and control of the environment and of homeostasis, leaving room for a notion which has been rather neglected, i.e. the adaptational capacities of the animal. Furthermore, welfare has too often been seen in the light of misleading concepts such as "behavioural needs" which became static and permanent points of reference. This simplistic misuse of ethology has been challenged a.o. by Dawkins (1983) who proposes the use of concepts from economical sciences such as "elastic" and "inelastic demands".

This paper re-evaluates experiments, analyzes litigious points and gives hints for future research in the light of Wiepkema's approach. I hope that up-to-now contradictory results will be better understood. I will allow myself to use data from other than farm animals when it contributes to understand a point.

When no alternative response can diminish the mismatch, cognitive psychologists will tell us that humans will change the norm, minimalize or negate that mismatch (e.g. Berlyne, 1960; Festinger, 1980). Animals can change important learned norms (Wiepkema,

1980); it seems less likely, however, that they are able to flee in an imaginary world, to rationalize or repress the mismatch, like humans can with their developed cortex. So they remain stuck with the problem on the motor level.

Two types of individual reactions can be observed (Koolhaas et al., 1983). Some individuals do not show any particular activity and it would be interesting to measure accurately if they show a real decrease in activity, which would indicate to what extent the Behavioural Inhibition System (B.I.S.) is still functioning (Gray, 1982; Laborit, 1974). Others develop various types of "abnormal behaviours" and continue to perform them, even if they are not useful to resolve the mismatch, even if they are repetitive, even if (sometimes) the motor pattern is not related to the usual motor repertoire available to reduce the mismatch in question (e.g. the so-called "displacement activities" and "low threshold" behaviours). It is as if in some individuals a programme is present inducing the animal to do something to act. It has been reported that rats in skinner-boxes receiving a reward without having to perform any particular behaviour, do not "accept" such free food and develop "superstitious behaviours" as if they had to do something to get it (Davis and Hubbard, 1972). Waiting while doing nothing seems to be a difficult thing to do!

Abnormal behaviours observed by ethologists in conflict situations, eventually experimentally induced, can be observed in farm animals in intensive management system. Several authors have catalogued these extensively (e.g. Brummer, 1978; Sambraus, 1985a; Vestegaard, 1984; Wiepkema et al., 1983).

They can be briefly categorized as redirection activities, stereotypies and apathy.

REDIRECTION ACTIVITIES

These activities can somethimes be (auto)mutilating and hence called "injurious behaviours". Blokhuis and Arkes (1984) presented good evidence that feather-pecking in chickens evolves as redirected food-pecking. One group of chickens was housed on litter, the other on slatted floor. Pecking at conspecifics increased with time in the slatted floor group, while it decreased in the litter group. Then, half of the animals from each group was transferred to the other floor type. The animals reared on litter and switched to slats increased feather-pecking with time. Floor-pecking was always high in groups with a low frequency of feather-pecking and vice-versa. The increase in rooting and biting at objects just before farrowing in sows could be due to the inability to perform normal nestbuilding (Baxter, 1982; Lammers and de Lange, 1986). Ladewig et al. (1984) observed that several sows kept on slats developed vulvar lesions caused by biting by the piglets. They subsequently discovered that piglets reared on slatted floors spent more time investigating various parts of the sow at 3 and 6 weeks of age than piglets reared on straw, while the latter spent more time investigating the straw than the former rooting the floor. De Wilt (1985) found that none out of 32 teat-bucket reared veal calves engaged in preputial sucking, while 24 out of 50 bucket-reared and 5 out of 15 cow-reared calves did so, which shows that early sucking experience - or the lack of it - influences the development of that undesirable behaviour. Several authors have observed that piglets reared in barren farrowing crates spend more time massaging and nibbling littermates or the sow than piglets in straw pens (Fraser, 1978; Schmidt and Adler, 1981; Schouten, 1986; van Putten and Dammers, 1976). Tail-biting is more frequent in pens with the least possibilities of rooting (Ruiterkamp, 1985).

Although the decrease of welfare of the eventual victims of injurious redirected activities is evident, there is still a controversy concerning the welfare of the actors. Discussions of that problem (and that of "vacuum activities") can be found in Dawkins (1980, 1983) and Sambraus (1981).

STEREOTYPIES

More or less extended discussions of stereotypies (SS) elicited by intensive captivity have been written by Broom (1983), Kiley-Worthington (1977), Ödberg (1978, 1981) and Sambraus (1985b). There exists a substantial literature on farm animals demonstrating that the more the environment seems indequate to the animal, the more likely SS will appear. Most of these studies do not concentrate on the mechanisms of SS, but use them as welfare indicators to evaluate zootechnical life conditions.

Few authors have experimented with SS as such. A.o., Duncan and Wood-Gush (1972) demonstrated that in fowl SS develop in function of intensity of frustration. Rushen (1985) suggests that frustrated feeding is the origin of the SS appearing in tethered sows before and after the actual feeding period, and that the different stereoptyic (S) patterns are derived from parts of the appetitive and consummatory phases. Additional food (=1/2h to 1h longer feeding times) and spreading of sawdust reduced sham chewing in individually penned sows, while bar- gnawing was only reduced when the latter was present (Sambraus and Schunke, 1982). Keiper (1969) could reduce independently route-tracing in canaries by increasing place, and spot-pecking by forcing the birds to work more to get food. Ödberg (1981), using bank voles as a cheap laboratory model, found that enrichement of the environment is more important for keeping down the amount of animals presenting SS than cage size. These results show that 1º a welfare legislation should not always be based on space only, 2º a careful analysis of the pattern of the abnormal behaviour shown can sometimes indicate, which aspects of the management or environment are inadequate.

One must, however, remember that when one has not followed the ontogeny of the SS and, ideally, provoked them experimentally, it may become very difficult to distinguish between the original causal factor and merely modulating ones such as arousing stimuli (human presence, feeding time, elimination, etc.) or the provision of alternative activities including those induced by novelty. An example out of the zoo-world will illustrate this. The S-level of fennecs dropped each time sand was provided during about one week; however, sand lost its effect when it remained for 3 months (Ödberg, 1986a). Cronin and Wiepkema (1984) observed in sows that the time spent performing SS increased with pregnancy, but dropped after the 80th day. At that stage hormonal changes occur in the sow which could have modulated the S-level.

A complicating fact is that SS are not simply performed when all other behaviours stop. They are subjected to the same factors which determine activity rhythms. E.g. tethered sows do not show them during the night (Cronin, 1985). Bank voles have a polyphasic activity rhythm with peaks of 1 to 4 hours every 1 to 6 hour. SS appear only during these peaks and can represent a substantial part of them (Ödberg, 1986b).

Furthermore, the neurobiochemical background of SS changes probably during their development. Duncan and Wood-Gush (1974) were able to inhibit SS in fowl with Pacitran (a Rauwolfia derivate), when given during the initial frustrating situation; the drug did not work any more once the SS were well established. Similarly, Feldman (1962) could refrain the development of "fixated choices" in rats submitted to an insoluble task in a Lashley stand by giving chlordiazepoxide, but the drug had no effect once these fixations were established. Cronin et al. (1985) found that naloxone inhibited more the newer SS than the older ones in tethered sows. A systematic study of this phenomenon is presently being carried out in Ghent.

For all these reasons, little can be discovered about the mechanisms of SS as long as one does not carry out a developmental study. Or at least, one must know the "age" of the SS one is working with.

Stolba et al. (1983) and Cronin (1985) have observed sows from the moment of tethering on. In a first stage (data from Cronin), The animals reacted vigorously with escape attempts and aggressive acts (average duration 45 min). In stage 2, they remained lying idle and immobile for an average of 1 day. Stage 3 was characterized by an increased performance of various behaviours, the reappearance of initial conflict behaviours and the first development of SS (average 16 days). Stage 4 consisted of up to 70% of a sow's activity during the observation time being devoted to the performance of various idiosyncratic SS. The first part of this evolution fits easily into the cybernetical model. The animal first reacts by a succession of responses of which none can reduce successfully the mismatch (i.e. free oneself) (stage 1); the

B.I.S. dominates and inhibits the planning of new motor programmes (stage 2) and the animal is left without answers in a total lack of control of his own life. But then? As not to act is unbearable, the animal starts again doing something (stage 3) and some movements must be rewarded in some way so that they are repeated, even without reducing the original mismatch. What is then reduced? Many authors have suggested that SS do have a function, i.e. to reduce arousal, anxiety or frustration, whatever one calls it, but without presenting evidence. One cue was found by Duncan and Wood-Gush (1972a) who observed that hungry chickens unable to reach food started to emit alarm-calls at a given frustration level; these calls disappeared when they had developed a high frequency of stereotyped pacing. As it is known that endorphins, that are chemically related to ACTH (Guillemin et al., 1977) have analgesic and quietening effects (Oliverio et al. 1984; Watkins and Mayer, 1982), Cronin et al. (1985, 1986) injected sows with the endorphin receptor-blocker naloxone. This drug inhibited spectacularly the SS, the interesting point being that some sows started to show behaviours typical for stage 1. The final demonstration would be a follow-up study in which the level of some neuroendocrine stress parameters would decrease in opposite relation with the devlopment of SS. Such a study is now being undertaken in Ghent. In the meanwhile, Dantzer and Mormède (1983b) have found that post-session corticosteroid levels were lower than pre-session levels, when pigs submitted to an intermittent food delivery schedule could engage in stereotyped manipulation of a chain. A puzzling aspect of the results, however, was that this difference was due to increased pre-session levels in the pigs which could manipulate the chain as compared to those which could not.

PUTATIVE FUNCTIONS

These data hint that SS are indeed a homeostatic mechanism to reduce arousal. Delius (1967), experimenting with electrical brain stimulation in gulls, had already suggested such a function for some displacement activities, based on the discovery that behaviours often appearing as displacement activities (such as grooming and preening) were largely controlled by neurophysiological mechanisms also responsible for de-arousal and sleep. As early as 1940, Tinbergen observed that displacement activities were usually incompletely performed. Duncan and Wood-Gush (1972b) demonstrated with the help of frame to frame film analyses, that displacement preening in chickens is quantitatively and qualitatively different from normal preening. Hence, there must be something more than just an attention switch (McFarland, 1966) to other stimuli, aimed at the release of any alternative response reducing effectively the mismatch.

The motor pattern used as stereotypy can sometimes be traced back to an intention movement belonging to the behavioural repertoire of the frustrated motivation (cf. Ödberg, 1978), but most of the time it belongs to those behaviours also often observed as displacement activities. Noteworthy is that the arousal-reducing effect could be associated with one stereotypy and not with the other, eventually within the same individual animal (cf. Wiepkema on veal calves, this volume). Puzzling was also Kruijt's (1964) observation on fighting Bankiva cocks: some displacement activities were more often observed in winners, others more in losers. The motor pattern can further be submitted to a simplification process. Its frequency can depend less and less from the actual environmental stimuli, until it becomes totally emancipated from the original causal situation. This has partly been observed in the development of the SS in tethered sows (Cronin, 1985; Stolba et al., 1983). At that moment, the chances of reducing the original mismatch as such seem completely lost. One must be dealing here with a different and particular function.

Cues are now coming in about the consequences of that function. SS may benefit to the organism by protecting it from psychosomatic illnesses. Veal calves performing SS had less or no abomasal lesions as compared to those which did not. More data about the biological benefit of SS can be found in Wiepkema's preceding paper.

Let us mention at once that even if SS are a coping mechanism, we should not infer that welfare is assured; it is very likely that the animal has been through a very nasty period before and during the earlier stages of the development of the SS, as we have seen with the tethered sows. Furthermore, as Broom pointed out (this volume), paying a high cost in order to cope means also a decreased welfare. Little is known about this up to now. Cronin et al. (1986) found that the performance of high levels of SS are indeed associated with increased metabolic rate.

INDIVIDUAL

If annoying to the statistician, individual differences or exceptions can be very interesting to the ethologist. Why do some animals develop SS and others not?

Large individual differences in S-level in sows have been recorded (Bengtsson and Svendsen, 1983; Barnett et al., 1985; Cronin, 1985, p.94). Bareham (1972) found a positive correlation between S-level and adrenal weight in fowl. This measurement was carried out soon after the initiation of SS and would indicate that the most stressed birds performed the most SS. Sacrifying the animals at different stages in the S-development could have yielded even more revealing data; would the adrenal weight have subsided with the further development of SS and remained high in those individuals which did not keep a high S-level? Cabriolet and Dantzer (1984) found that in tethered sows hyperactivity was associated with high S-levels. Similarly, Ödberg (1986b) observed that the more active voles tended to show higher S-levels and vice-versa. Once again, the snag in these studies is that they describe a given situation at a given period. In a follow-up study, however, Ödberg (unpublished results) found initially that the performance of SS by voles was significantly associated with higher levels of walking-sniffing and of rearing, while a negative relation was found with immobility (sleeping excluded). When animals stopped to perform SS, either spontaneously, or because of being transferred from a barren environment to an enriched one, the time spent walking-sniffing and rearing decreased and immobility increased. The reverse trend was found in the voles which pursued the opposite evolution (i.e. which began to show SS). All these data suggest that SS are a dynamic response to frustration - the more active animals being more frustrated - and not a "compensation" according to a rule which would state that an organism should be active a given percentage of the time. In the latter case a negative correlation should have been found between general activity and S-level.

The study of individuals which do not develop abnormal behaviours could yield data very interesting to welfare students.

The following hypotheses can be set up concerning these individuals.

1. They simply do not experience the important mismatches that others do; this implies they would be endowed with quite different norms; this seems unlikely to me.
2. They are able to reduce better the mismatch
   A. either because they have less stringent norms, or
   B. because they can implement equally or more efficient strategies.
3. They experience the mismatch frustration as much as the others do, but cannot develop any coping mechanism, i.e. the incapacity to develop an abnormal behaviour leaves them in an even greater distress.

I hope the developmental study undertaken in Ghent will yield cues about to which group such individuals belong. Theoretically, corticosteroid levels could in general either remain low and behavioural observations could perhaps determine if these individuals belong to 2A or to 2B, or remain high (group 3). These considerations bring us to another possible reaction to chronic stress: apathy.

APATHY

It must first be pointed out that a lower activity level as such does not mean that the animal reacts less to stressors. Faure (1980) found that hens from an active strain had a lower basal plasma corticosterone level than hens from an inactive one, and that this difference increased when a stressor was given. Similar phenomena were observed in rats (Gentsch et al., 1982). The heart rate of hens from a placid strain rose nearly as high as that of a flighty strain in reaction to a visual stressor, but it remained high much longer (Duncan and Filshie, 1979). One can wonder whether active strains are active and less stress-susceptible, and whether they show lower stress parameters because their higher activity has de-arousing properties. A point which should be heeded by those engaged in genetic selection for animals better adapted to intensive management (for discussions on this topic see: Beilharz, 1982; Faure, 1980; van Rooyen, 1983).

Until recently, the interpretation of lasting immobility as an abnormal reaction has not been substantiated scientifically in farm animals. Sambraus and Schunke (1982) suggest that in tethered sows it is rather a direct reaction to cold surfaces and unadequate box size. Great differences between housing systems for sows have not been found (Barnett et al., 1984) and when some were found, longer immobility periods could have been also directly due to physical restraint. Broom (1986), however, found lately that stall-housed sows reacted remarkably less to water poured into their backs than group-housed ones (see also Broom this volume). The former laid also longer motionless and performed a lesser variety of behaviours.

A difficulty is the exact description of apathy. An immobile animal can be sleeping, resting, freezing, simply watching, either relaxed or intensively seeking stimuli. All this is different from apathy. One could furthermore wonder if learned helplessness has something in common with apathy. Recent reports suggest that, as SS, the former may reflect a conditioned analgesic response, but mediated by enkephalins (for a review, see Gray, 1982, pp.385-387).

BEHAVIOURAL DE-AROUSAL AND ENDOCRINOLOGICAL WELFARE ASSESSMENT

Although a number of studies show a positive relation between physiological measures of pituitary-adrenal activity and the occurrence of abnormal behaviours (e.g. Stuhec, 1984; Schlichting, 1984), others yielded puzzling results or apparently contradictory ones (Barnett et al., 1984, 1985; Becker et al., 1984).

One must of course realize that little is still known about the effects of chronic stress on the neuroendocrine system of farm animals (Mormède et al., 1983) and that one should look for new indicators (Dantzer et al., 1983). However, the homeostatic function of abnormal behaviours should also be taken into account; maybe it is responsible for puzzling results. Neurophysiological parameters can vary according to many variables such as diurnal rhythm (Ladewig, 1984), genetic background (breeds, strains, H^+/H^-) (Craig, 1970; Duncan and Filshie, 1979; Faure, 1980; Mormède and Dantzer, 1978), imprinting and early learning (Denenberg, 1969), hierarchical status (Stuhec, 1984) homeostatic mechanisms at the neurobiochemical level (Burchfield et al., 1980; Moss, 1981; Owen et al., 1981). One is thus bound to carry out follow-ups of physiological and behavioural parameters simultaneously in individual animals , i.e. to discover what changes within the animal and how (e.g. Beuving, 1983; Moss, 1981).

Furthermore, one has to take into account the notion of "role", i.e. that each animal plays al given role in a given society, at a given age, with a given sex and a given reproductive status, and could have fluctuating norms in function of that role. This implies that one should choose carefully which parameter relevant for each role one will measure.

REFERENCES

Bareham, J.R. 1972. Effect of cages and semi-intensive deeplitter pens on the behaviour, adrenal response and production in two strains of laying hens. Br. Vet. 9, 153-163.

Barnett, J.L., Cronin, G.M., Winfield, C.g. and Dewar, A.M.., 1984. The welfare of adult pigs: the effect of five housing treatments on behaviour, plasma corticosteroids and injuries. Appl. Anim. Behav. Sc. 12, 209-232.

Barnett, J.L., Winfield, C.G., Cronin, G.M., Hemsworth, P.H. and Dewar, A.M., 1985. The effect of individual and group housing on behavioural and physiological responses related to the welfare of pregnant pigs. Appl. Anim. Behav.Sc. 14, 149-161.

Baxter, M.R. 1982. The nesting behaviour of sows and its disturbance by confinement at farrowing. In "Disturbed behaviour in farm animals" (Ed. W. Bessei). (Hohenheimer Arbeiten 121, Stuttgart). pp. 101-114.

Becker, B., Christenson, R., Ford, J. Manak, R., Nienaber, J., Hahn, G. and Deshazer, J., 1984. Serum cortisol concentrations in gilts and sows housed in tether stalls, gestation stalls and individual pens. Ann. Rech. Vét. 15, 237-242.

Beilharz, R.G. 1982. Genetic adaptation in relation to animal welfare. Int.J.Stud.Anim.Prob. 3, 117-124.

Bengtsson, A.C. and Svendsen, J. 1983. Housing of sows in gestation. 34th Ann.Meeting EAAP, Madrid. p.398.

Berlyne, D.E. 1960. "Conflict, Arousal and Curiosity", McGraw-Hill, New York.

Beuving, G. 1983. Corticosteroids in welfare research of laying hens. In "Indicators Relevant to Farm Animal Welfare" (Ed.D. Smidt). (Martinus Nijhoff for the CEC, Boston). pp. 47-53.

Blokhuis, H.J. and Arkes, J.G. 1984. Some observations on the develoment of feather-pecking in poultry. Appl.Anim.Behav.Sc. 12, 145-157.

Broom, D. 1983. Stereotypies as animal welfare indicators. In "Indicators Relevant to Farm Animal Welfare" (Ed. D. Smidt). (Martinus Nijhoff for the CEC, Boston). pp. 81-87.

Broom, D. 1987. Responsiveness of stall-housed sows. Appl.Anim. Behav.Sc. (in press).

Brummer, H. 1978. Verhaltensstörungen. In "Nutztierethologie" (Ed. H.H. Sambraus). (P. Parey, Berlin & Hamburg). pp. 281-292.

Burchfield, S.R., Woods, S.C. and Elich, M.S., 1980. Pituitary adrenocortical response to chronic intermittent stress. Physiol. Behav. 24, 297-302.

Cariolet, R. and Dantzer, R. 1984. Motor activity of pregnant tethered sows. Ann. rech. Vét. 15, 257-261.

Craig, J.V. 1970. Interactions of genotypes and housing environment in white Leghorm chickens selected for high and low social dominance. Proc. 14th World's Poultry Congress, Madrid, Vol. 2, 37-42.

Cronin, G.M. 1985. "The Development and Significance of Abnormal Stereotyped Behaviours in Tethered Sows". Ph.D. thesis Agricultural University Wageningen.

Cronin, G.M., van Tartwijk, J.M.F.M., van der Hel, W. and Verstegen, M.W.A. 1986. The influence of degree of adaptation to tether-housing by sows in relation to behaviour and energy metabolism. Anim. Prod. 42, 257-268.

Cronin, G.M. and Wiepkema, P.R. 1984. An analysis of stereotyped behaviour in tethered sows. Ann. Rech.Vét. 15, 263-270.

Cronin, G.M., Wiepkema, P.R. and van Ree, J.M. 1985. Endogenous opioids are involved in abnormal stereotyped behaviours of tethered sows. Neuropeptides 6, 527-530

Cronin, G.M., Wiepkema, P.R. and van Ree, J.M. 1986. Endorphins implicated in stereotypies of tethered sows. Experientia 42, 198-199.

Dantzer, R. and Mormède, P. 1983a. Stress in farm animals: a need for reevaluation. J. Anim.Sc. 57, 6-18.

Dantzer, R. and Mormède, P. 1983b. De-arousal properties of stereotyped behaviour: evidence from pituitary-adrenal correlates in pigs. Appl.Anim.Ethol. 10, 233-244.

Dantzer, R., Mormède, P. and Henry, J.P. 1983. Significance of physiological criteria in assessing animal welfare. In "Indicators Relevant to Farm Animal Welfare" (Ed.D.Smidt). (Martinus Nijhoff for the CEC, Boston). pp.29-37.

Davis, H. and Hubbard, J. 1972. An analysis fo superstitious behavior in the rat. Behaviour 43, 1-12.

Dawkins, M.S. 1980. "Animal Suffering. The Science of Animal Welfare". (Chapman & Hall, London, New York).

Dawkins, M.S. 1983. Battery hens name their price: consumer demand theory and the measurement of ethological "needs". Anim.Behav. 31, 1195-1205.

Delius, J.D. 1967. Displacement activities and arousal. Nature 214, 1259-1260.

Denenberg, V.H. 1969. The effects of early experience. In "The Behaviour of Domestic Animals" (Ed. E.S.E. Hafez). (Bailliere, Tindall & Cassell, London). pp. 95-130.

De Wilt, J.G. 1985. "Behaviour and Welfare of Veal Calves in Relation to Husbandry Systems". Ph.D.Thesis, Agricultural University Wageningen.

Duncan, I.J.H. and Filshie, J.H. 1979. The use of radio telemetry devices to measure temperature and heart rate in domestic fowl. In "A Handbook on Biotelemetry and Radio Tracking" (Eds. C.J. Amlaner & D.W. Macdonald). Pergamon Press, Oxford). pp. 579-588.

Duncan, I.J.H. and Wood-Gush, D.G.M. 1972a. Thwarting of feeding behaviour in the domestic fowl. Anim. Behav. 20, 444-451.

Duncan, I.J.H. and Wood-Gush, D.G.M. 1972b. An analysis of displacement preening in the domestic fowl. Anim. Behav. 20, 68-71.

Duncan, I.J.H., and Wood-Gush, D.G.M. 1974. The effect of a rauwolfia tranquillizer on stereotyped movements in frustrated domestic fowl. Appl.Anim.Ethol. 1, 67-76.

Faure, J.M. 1980. To adapt the environment to the bird or the bird to the environment? in "The Laying Hen and its Environment" (Ed.R.Moss). (Martinus Nijhoff for the CEC, the Hague). pp. 19-30.

Feldman, R.S. 1962. The prevention of fixations with chlordiazepoxide. J. Neuropsychiat. 3, 154-259.

Festinger, L. 1980. Cognitive dissonance. In "Mind and Behaviour" (Readings from Scientific American). (W.H. Freeman & Co). pp. 267-273.

Fraser, D. 1978. Observations on the behavioural development of suckling and early-weaned piglets during the first six weeks after birth. Anim.Behav. 26, 22-30.

Friend, T.H. 1980. Stress: what is it and how can it be quantified? Int.J.Stud.Anim.Prob. 1, 366-374.

Gentsch, C., Lichtensteiner, M., Driscoll, P. and Feer, H. 1982. Differential hormonal and physiological response to stress in Roman high- and low-avoidance rats. Physiol.Behav. 28, 259-263.

Gray, J.A. 1982. "The Neuropsychology of Anxiety". (Clarendon Press, Oxford, Oxford University Press, New York).

Guillemin, R., Vargo, T., Rossier, J., Minick, S., Ling, N., Rivier, J., Vale, W. and Bloom, F. 1977. Beta-endorphin and adrenocorticotropin are secreted concomitantly by the pituitary. Science 197, 1367-1369.

Keiper, R.R. 1969. Causal factors of stereotypes in caged birds. Anim.Behav. 17, 114-119.

Kiley-Worthington, M. 1977. "Behavioural Problems of Farm Animals". (Oriel Press, Stocksfield).

Koolhaas, J.M., Schuurman, T. and Fokkema, D.S. 1983. Social behaviour of rats as a model for the psychophysiology of hypertension. In "Biobehavioural Bases of Coronary heart Disease" (Eds. T.M. Dembrowski, Schmidt, T.H. & Blümchen, G.). (Karger, Basel). pp. 391-400.

Kruijt, J.P. 1964. Ontogeny of social behaviour in Burmese Red Junglefowl (Gallus gallus spadiceus Bonaterre). Behaviour Suppl. N°12.

Laborit, H. 1974. Action et reaction. Mecanismes bio- et neurophysiologiques. Aggressologie 15, 303-322.

Ladewig, J. 1984. The effect of behavioural stress on the episodic release and circadian variation of cortisol in bulls. Proc. Int.Congress on Appl.Ethol. in Farm Anim., Kiel. pp 339-342.

Ladewig, J., Klöppel, P. and Kallweit, E. 1984. A case of "reversed cannibalism": the piglets damaging the sow. Ann Rech.Vét. 15, 275-277.

Lammers, G.J. and de Lange, A. 1986. Pre- and post-farrowing behaviour in primiparous domesticated pigs. Appl.Anim.Behav. Sc. 15, 31-43.

McBride, G. and Craig, J.V. 1985. Environmental design and its evaluation for intensively housed animals. Appl.Anim.Behav. Sc. 14, 211-224.

McFarland, D.J. 1966. On the causal and functional significance of displacement activities. Z. Tierpsychol. 23, 217-235.

Mormède, P. and Dantzer, R. 1978. Behavioural and pituitary-adrenal characteristics of pigs differing by their susceptibility to the malignant hyperthermia syndrome induced by halothane anesthesia. 2. Pituitary-adrenal function. Ann.Rech. Vét. 9, 569-576.

Mormède, P., Bluthé, R.M. and Dantzer, R. 1983. Neuroendocrine strategies for assessing welfare: application to calf management systems. In "Indicators Relevant to Farm Animal Welfare" (Ed.D.Smidt). (Martinus Nijhoff for the CEC, Boston). pp. 39-46.

Moss, B.W. 1981. The development of a blood profile for stress assessment. In "Welfare of Pigs" (Ed.W.Sybesma). (Martinus Nijhof for the CEC, The Hague). pp 112-125.

Ödberg, F.O. 1978. Abnormal behaviours: stereotypies. Proc.Ist. World Congr. on Ethol. applied to Zootechn. Madrid. pp. 475-480.

Ödberg, F.O. 1981. "Environmental and pharmacological determinants of pathological stereotypies in voles (Clethrionomys glareolus britannicus)". Ph.D. Thesis, State University of Ghent. (in Dutch).

Ödberg, F.O. 1986a. De modulation of stereotypy levels in fennecs (Fennecus zerda) by the mediation of sand. Acta Zool.Pathol. Antv. (in press).

Ödberg, F.O. 1986b. The jumping-stereotypy in the bank vole. Biol.Behav. 11, 130-143.

Oliverio, A., Castellano, C. and Puglisi-Allegra, S. 1984. Psychobiology of opioids. Int.Rev.Neurobiol. 15, 277-337.

Owen, F., Baker, H.F., Ridley, R.M., Cross, A.J. and Crow, T.J. 1981. Effect of chronic amphetamine administration on central dopaminergic mechanisms in the vervet. Psychopharmacologia 74, 213-216.

Ruiterkamp, W.A. 1985. "Het gedrag van mestvarkens in relatie tot huisvesting". Ph.D. Thesis, State University of Utrecht. (with English summary).
Rushen, J.P. 1985. Stereotypies, aggression and the feeding schedules of tethered sows. Appl.Anim.Behav.Sc. 14, 137-147.
Sambraus, H.H. 1981. Abnormal behaviour as an indication of immaterial suffering. Int.J.Stud.Anim.Prob. 2, 245-248.
Sambraus, H.H. 1985a. Mouth-based anomalous syndromes. In "Ethology of Farm Animals". (Ed.A.F.Fraser). (Elsevier, Amsterdam). pp. 391-422.
Sambraus, H.H. 1985b. Stereotypies. In: as Sambraus 1985a. pp. 431-441.
Sambraus, H.H. und Schunke, B. 1985. Verhaltensstörungen bei Zuchtsauen im Kastenstand. Wien. Tierärztl.Mschr. 69, 200-208.
Schlichting, M. 1984. Ethophysiological studies in gilts. Ann Rech.Vét. 15, 243-244.
Schmidt, M. and Adler, H.C. 1981. Danish studies on behaviour of early weaned piglets: preliminary results. In "The welfare of Pigs" (Ed.W.Sybesma). (Martinus Nijhoff for the CEC). pp. 211-223.
Schouten, W.G.P. 1986. "Rearing conditions and behaviour in pigs". Ph.D. Thesis, Agricultural University Wageningen.
Stolba, A., Baker, N. and Wood-Gush, D.G.M. 1983. The characterisation of stereotyped behaviour in stalled sows by informational redundancy. Behaviour 87, 157-182.
Stuhec, I. 1984. "Ethologische und verhaltensphysiologische Untersuchungen zur Belastung von Jungsauen durch verschiedene Haltungssysteme". Ph.D. Thesis, Universität zu Kiel.
Tinbergen, N. 1940. Die Übersprungbewegung. Z. Tierpsychol. 4, 1-40.
Van Putten, G. and Dammers, J. 1976. A comparative study of the well- being of piglets reared conventionally and in cages. Appl.Anim. Ethol. 2, 339-356.
Van Rooijen, J. 1983. Genetic adaptation and welfare. Int.J.Stud. Anim. Prob. 4, 191-197.
Vestergaard, K. 1984. An evaluation of ethological criteria and methods in the assessment of wellbeing in sows. Ann.Rech. Vet. 15. 227-235.
Watkins, L.R. and Mayer, D.J. 1982. Organisation of endogenous opiate and non-opiate pain control systems. Science 216, 1185-1192.
Wiepkema, P.R. 1980. Gedrag en welzijn. Inaugural speech, Agricultural University Wageningen.
Wiepkema, P.R., Broom, D.M., Duncan, I.J.H. and van Putten, G. 1983. Abnormal behaviour in farm animals. Report of the commission of the European Comunities.

INTRODUCTION

During this workshop it was repeatedly stated that we only can advance our insight into interrelated concepts like adaptation, stress and animal welfare, substantially, if we succeed in integrating various disciplines in one and the same model of animal functioning. At this moment such a model is still far away. Although this state of affairs will have to do with serious gaps in our knowledge of basic mechanisms in, for instance, neurobiology, it certainly also results from our specialist's attitude in science. For instance, the more fascinating a discovery on the molecular level, the greater the temptation to overlook the intact individual. However, there is a growing need to return to the whole animal. This return is indeed a must in so far we want to make reliable and interesting statements about concepts like animal welfare and stress. Toates' chapter represents first steps on this way by bringing together a set of relevant topics out of present day research on animal behaviour. By reviewing and comparing relevant theories about the role of motivation in animal behaviour, he makes clear that at least in higher vertebrates we have to skip the idea that these animals are little more than simple stimulus-response systems. Higher vertebrates not only react, they also act and take initiatives: basically these animals are flexible goal directed systems.

If we accept this, and Toates' arguments are strong indeed, we should also make explicit as well as possible the consequences of such an assumption. Toates tries to find a perspective by discussing the essentials of present day theories on motivation and learning. His conclusion is that higher vertebrates have Sollwerte and expectancies. If we take this cognitive aspect seriously, it entails that individuals are able to interpret the outcome of their behaviour programs in such a way that this outcome may be qualified as successful, failing or something in between. In other words, the outcome of behaviour is not irrelevant; it has an individually determined value. For that reason motivated behaviour is often affect loaded, as put forward by Epstein (cited by Toates). It is on this point that the contribution of Cabanac comes in.

Cabanac argues that individual behaviour can best be described as aimed at maximizing pleasure and minimizing displeasure. Of course, the actual solution will always be some compromise. The point emphasized by Cabanac is that in animals and men emotions are indispensable for shaping actual behaviour output. It is quite possible, however, that this relationship between (dis)pleasure and behaviour is only relevant as long as the behaviour involved is not yet a routine (cf. Wiepkema's chapter). Without doubt, we urgently need systematic and integrative research into the role of emotions in adaptive behaviour; in this the distinction between emotional expressions and emotional experiences will be quite relevant.

Toates' discussion of motivational and learning processes in goal directed systems brings him imperceptibly to the two keywords of present day stress research: 1) controllability and 2) predictability of environmental changes. Basically stress appears to result from an informational deficit, that typically may arise in organisms characterized by cognitive brain processes. Loss of controllability or of predictability evokes expressions of negative emotions and even may lead to significant signs of helplessness as manifested by a generalized anxiety, a decreased tendency to react to aversive stimuli and a loss of learning capability.

The existing evidence induces Toates to conclude that certainly higher vertebrates, like our farm animals, really may be stressed and suffer as a result of the housing conditions they live in. As has been put forward in foregoing chapters this may have far reaching consequences for these animals in terms of health and welfare, of production and of reproduction.

THE RELEVANCE OF MODELS OF MOTIVATION AND LEARNING TO ANIMAL WELFARE

Frederick Toates
Biology Department,
The Open University,
Milton Keynes,
United Kingdom.

ABSTRACT

A model of motivation and learning, in which the organism is viewed as goal-directed and capable of learning cognitions, is described. The relevance of this model for animal welfare and stress reduction is outlined, in terms of the animal being allowed to attain the goals that it sets. Older S-R models are reviewed, and evidence presented to show their inadequacies. The importance of allowing the animal the opportunity to perform species-typical behaviour under certain circumstances is discussed.

INTRODUCTION

Background

At any point in time, attitudes to animal welfare, suffering and husbandry might be expected to be influenced by the models and theories that we employ as descriptions and explanations in the behavioural sciences. Of course, common sense has always been able to tell us much about animal welfare. What a scientific understanding can add that is of value, over and above what we would deduce from common sense and a feeling of empathy, is only as good as the models of behaviour that are currently in fashion. For example, a Cartesian 'model' involving clear dichotomies (a) between mind/soul and brain and (b) between humans and brutes, might lead one to suppose that suffering, with its conscious affective aspect, is an exclusively human phenomenon. Fortunately, we have moved on some way from such a model. This chapter critically reviews some theories and models of behaviour, pointing to their strengths and weaknesses. I shall concentrate upon past and present models of motivation, learning and goal-directed behaviour, showing the interdependence between these aspects. I shall summarize what I perceive to be the contemporary Zeitgeist regarding behavioural models and indicate what I interpret as being its essential message, as far as the issue of animal welfare is concerned.

Some tentative preliminary pointers as to the relevance of behavioural models to animal husbandry will be discussed, and it is hoped that a more sound ethological basis for future decisions can be given. Further, there is the possibility of a two-way flow of information: consideration of welfare issues might illuminate pure behavioural science.

Definitions and domain

Under what circumstances might we be inclined to say that an animal is suffering? If it is organically ill or subject to physical injury, we would logically assume that it is suffering. Controversy is more likely to arise if we were to propose that an animal is suffering as a result of, say, isolation from conspecifics or overcrowding. Such suffering is often described under the heading of 'stress'. Reflecting this, most of what follows is indeed on the subject of stress. However, I shall also discuss more general ways in which behaviour models can illuminate welfare issues.

From the viewpoint of stress, a useful way of approaching the question of suffering, and one that will be employed here, was outlined by Dawkins (1980). She starts her argument on the premise that there is a 'stress syndrome' consisting of hormonal and other physiological changes that can ultimately lead to organic damage. This syndrome appears following the presentation of a variety of stressors. Dawkins suggests that we can employ indices of stress that are apparent before organic damage results, in order to recognize and take steps to preempt suffering.

There is some controversy over exactly how we define stress (Dawkins, 1980). Archer (1979) defines it as "the prolonged inability to remove a source of potential danger, leading to activation of systems for coping with danger beyond their range of maximal efficiency." Archer notes the essential negative feedback nature of the system for dealing with threats. Threats cause (a) activation of behavioural systems, e.g. fear and attack, and (b) hormonal changes. In turn, these normally lead to removal of the source of threat. In stress situations, one has essentially opened the feedback loop so that the natural consequences (distancing of the threat) do not follow from the effect induced. As Broom (1985) notes, short-term adrenal activation, accompanied by appropriate behaviour should be seen as being of adaptive value to the animal. However, prolonged hyperactivity of the adrenal hormone system impairs, rather than helps, survival and reproduction. Broom distinguishes between adrenal activity of a magnitude that can help an animal to cope with the demands of the environment and harmful adrenal responses. In a similar vein to Archer, Broom (1985) defines stress as "the process by which environmental factors over-tax control systems in an individual thus activating responses whose effects are prolonged and ultimately detrimental to that individual..." Indicators of stress in poultry include increased injury, greater than normal susceptibility to illness, and decreased growth rate.

An analogy with the drinking motivation system might help illuminate stress. Clearly, prolonged dehydration is undesirable, will lead to tissue damage, and can be assumed to be aversive. However, mild degrees of dehydration will goad the animal to seek water, and play a vital role in its survival. There is little reason to suppose that, so long as water is reasonably accessible, mild dehydration is aversive (a similar argument in the context of a different analogy is given by Dawkins (1980)).

Clearly, the response to different stressors will differ in some respects. However, for the concept of stress to be a useful one we need to see how far we can apply a generic model. One might suppose that some stimuli are able to elicit stress by their intrinsic properties, and could be termed unconditional stressors (e.g. loud sounds, electric shock). Although the distinction is far from absolute, other stimuli might be termed conditional stressors; their value would depend upon what they signal. Regarding the first class, Katz (1983) notes that by classifying extremes of heat and cold, shock, trauma and loud noise as 'stressors' we can turn to their common property that they all evoke <u>escape</u> responses. Psychologically effective stressors might then be defined as also those having this property. In other words, in the terms used here, within its biological constraints, an animal will perform <u>goal-directed activity</u> to alleviate a stressful situation. Such a notion can give us a unifying concept applying between different stressors and cutting across any dichotomy between so-called 'physical' and 'psychological' stressors. For example, a rat can take evasive action to avoid heat or cold, and a duckling can perform an operant to reestablish contact with a parent on whom it is imprinted. A rat will escape from a goal-box in which it experiences frustrative non-reward (Archer, 1976).

The other aspect that might sometimes help sharpen our notion of what is a stressor would be to see it in terms of something _imposed_ upon the animal. Thus Katz notes that an animal might voluntarily separate itself from its family, e.g. during primate individuation. This should be placed in distinction to the stress of enforced separation. Katz continues: "Most stimuli allow preparation and subsequent stress minimization or escape. Inescapable stimuli (i.e. stressors) do not allow either of the last responses. They may force adaptation to occur exclusively at a motivational level, through the inhibition of normal affective responses. In conclusion, stress models by definition impose aversive stimulation and in so doing create motivational and behavioural changes as a direct consequence of limited escape options." On first exposure to inescapable shock, rats show escape behaviour, employing various strategies. A failure of such strategies is followed by 'giving up', inactivity and an absence of attempts to escape, characterized as 'helplessness' or 'hopelessness' (see discussion in Hellhammer, 1983).

Regarding possible conditional stressors, Levine, Weinberg and Ursin (1978) discuss the type of stimuli that can arouse either coping responses (see later) or stress, the response depending upon the animal's history and the environmental possibilities. In other words, the response depends upon the animal's _interpretation_ of the stimulus input. The same physical stimulus can evoke no signs of stress or major signs depending upon whether a coping response is available. This characteristic gives the stress system a feature in common with other motivational systems (see discussion in Toates, 1986). For example, a sucrose flavour might be avidly sought if it is predictive of benign consequences, i.e. energy yield. If the same physical stimulus of sucrose has been associated in the past with noxious consequences, it can evoke active avoidance reactions, the so-called _Garcia effect_.

Signs of animal discomfort

In this section, I shall very briefly summarize some of better known and more important cases of 'deviant' behaviour by farm animals. This section forms that material that is in need of explanation and prescription for improvement, and for which models of motivation and learning are relevant. It draws upon the review of Wiepkema (1985a).

Veal calves are normally housed in small wooden boxes (60-65 cm) that severely restrict their physical movements. In such an environment, they exhibit the bizarre habit of 'tongue playing', in which the tongue is rolled in and out or swayed from side to side. Veal calves housed in groups, but fed from a bucket, sometimes develop the habit of sucking each other, particularly just after milk drinking. The objects sucked include the ears, navel and penis of their fellows. In many cases, this leads to urine sucking and urine drinking.

Similarly, to save farm space, pregnant sows are sometimes held tethered with a band around their neck or breast. They are confined for long periods of time, and this precludes their performance of exploration. Sows typically develop a number of stereotypies, such as bar biting and sham-chewing (Cronin, 1985; see also Wiepkema, 1985a). Fattening pigs are commonly housed in pens without any bedding. This, of course, precludes such natural activities as rooting or nibbling at bedding. Rooting is then redirected, the target being penmates. Tail biting often appears and this can lead to severe wounds.

In the domain of poultry keeping, growing chicks and laying hens that are maintained without litter (e.g. on wire) exhibit pecking at each other. Such pecking is often associated with pulling of feathers and can cause serious injury. Laying hens, housed in battery cages, exhibit restless behaviour in the period just prior to laying. To Wiepkema (1985a):

"This pacing and putting the head out of the cage may last for months and strongly suggests an intense desire to escape from the enclosure."

Such laying hens also exhibit 'sham dust-bathing' on the grid floor of their cages. As an aspect of this behaviour, they attempt to rub their heads in the feed located outside their cages.

Wiepkema justifies defining these behaviours as 'abnormal', on the grounds that they are not shown under less confined, and more rich, natural environments. Furthermore, they are often to the animal's detriment (or to that of its fellows).

In the sections that follow, I shall attempt to show the relevance of various models and theories in psychology and ethology to such behaviours.

MODELS BASED UPON HOMOLOGY

Dawkins (1980) argues that we can logically infer the existence of mental states in animals. Hence, we can logically infer the existence of suffering.

A possible source of insight is to look at the similarities in the brains and behaviour of humans and non-human advanced animals. In this direction, Panksepp (1982) argues that there exist a few "hard-wired brain circuits" underlying the basic emotions and these are common in their structure between humans and non-humans. For example, in animals with ventromedial hypothalamic lesions, voraciousness and rage are highly probable, as is the case for humans. In contrast to the inter-specific similarity in mechanisms underlying emotion, the differences in cortical mechanisms are enormous (Panksepp, 1982). The emotive circuits underlie the animal's commerce with the external world, e.g. a positive emotion is associated with foraging and the negative emotion of fear is associated with putting a distance between the animal and the source of the fear. Furthermore, the emotive circuits mature earlier developmentally than do the more cognitive mechanisms. Taking a similar line to Panksepp, Jaynes (1982) argues:

"All mammals, including ourselves have a basic set of genetically organized affects, specific aptitudes to respond to characterizable classes of stimuli or events in certain characterizable ways."

Similarly, Wiepkema (1985b) argues that "man and most if not all other vertebrates experience emotions in a comparable way." This assumption is based upon the observation of the similarity of anatomical, physiological and endocrinological aspects across vertebrate species.

THE RELEVANCE OF BASIC MODELS OF LEARNING

Stimulus-response models

The history of explanation in the behavioural sciences, has been heavily influenced by models that see the organism as being essentially passive. That is to say, it simply reacts to. Indeed, the word 'stimulus' implies such a 'reaction to'. In the absence of such impinging stimuli, the organism would be assumed to be inert. Such a view doubtless owes much to Descartes' division of living systems into humans endowed with a spirit and non-human animals that are automatons.

The behaviourism of, or derived from, Watson and Pavlov is in accordance with such a model. Hull (1952) described how animals form stimulus-response associations in learning a task. Hull also saw animals as being driven to reduce their level of drive. Such models would presumably lead one to suppose that, in the absence of obvious tissue damage or stimuli that elicit fear, an animal should be 'happy', in so far as such a term would mean much in that context. We now know that the stimulus-response model is inadequate in important respects. In its place, contemporary theory sees the animal as being (a) intrinsically active (see discussion in Young, 1966), rather than passive, even in the absence of impinging stimuli, (b) goal-seeking, or, in other words, purposive (Toates, 1986), (c) flexible and (d) learning cognitions (Bolles, 1979; Dickinson, 1980, Mackintosh, 1983) and (e) exploratory (Archer and Birke, 1983). The contemporary picture demands a fundamental reassessment of what constitutes the minimum needs of a caged animal.

Contemporary learning theory

Contemporary learning theory replaces the notion of the animal as a collection of S-R associations with a more cognitive model. Animals are able to assimilate information about their environment in a way that is not tied to responses (see Walker, 1983 for a review). They can form predictive associations between two events in the external world or between, say, the taste of a substance and postingestive consequences (Dickinson, 1980; Mackintosh, 1983; see also the chapter by Wiepkema in this volume). An animal is also able to form predictive associations between its own behaviour and the consequences that follow (Bolles, 1972). Given the occurrence of an unexpected event, the animal's nervous system is programmed to search for a predictive cue, sometimes involving exploratory behaviour (discussed by Toates, 1983).

Learning theory emphasizes the existence of such affective states as 'hope', 'relief' and 'frustration' (Mackintosh, 1983). For example, presentation of a cue predictive of shock can have very similar, if not identical effects, to a cue predictive of 'no food' in a situation where food is expected. A cue predictive of 'no shock' in a situation where shock might otherwise be expected has a similar, if not identical effect, to a cue predictive of a positive hedonic event. Learning theorists postulate a reciprocal antagonism between positive and negative affective states (Mackintosh, 1983). On this point, despite some differences in basic assumptions, one can see a convergence of the models based upon learning theory and Panksepp's (1982) psychobiological model of the emotions. In the latter, based upon evidence from neurophysiology and psychopharmacology, Panksepp proposes reciprocal inhibition between antagonistic emotive circuits. Rather neatly, evidence from stress research dovetails here also. Measurements of pituitary-adrenal activity in rats under different circumstances fit such a model. On an instrumental task in a Skinner box, a decrease in the size of reward, relative to what is expected, causes a 'frustration effect' - an increase in pituitary adrenal activity (Goldman, Coover and Levine, 1973). Conversely, obtaining a reward better than expected causes what Goldman et al. term an 'elation effect' - a decrease in pituitary adrenal activity to below its basal level. Levine, Goldman and Coover (1972) relate these notions to the learning theory model of O.H. Mowrer, towards which so much of contemporary learning theory converges.

An addition to cognitive developments in learning theory is the recognition that there are strong biological constraints on learning (Bolles, 1979). That is to say, some associations are learned easily and others with difficulty, if at all. In applied ethology, any learning task presented to an animal (e.g. to earn its own food by operant means) would need to take such constraints into account.

In addition to the study of associations formed by animals, there is now considerable evidence to show that animals form complex cognitive representations of the spatial dimensions of their environment and the events within it (Menzel, 1978; Dawkins, 1980; O'Keefe and Nadel, 1978).

MODELS OF MOTIVATION

Energy models

For some years now, one of the more widely-discussed models in ethology has been that of Lorenz (1966, 1981). This is a model of motivation that is in what we might call the 'European ethological' tradition, as opposed to the more 'psychological' models that we associate with Anglo-American behavioural science. The essence of the Lorenzian model is that motivational-specific energy accumulates as a function of the time that elapses during which the animal does not perform the behaviour in question. Correspondingly, the stimulus adequate to elicit a behaviour decreases as a function of time. For example, Lorenz describes cases of animals being relatively discriminating in their choice of sexual partners after a short period of deprivation. As the length of the period increases, so the stimulus adequate to arouse mating falls. There are a number of distinct questions that we need to ask about this model:

(1) How adequate is the model in explaining and predicting behavioural phenomena?

(2) How general is the model? Does it apply to any motivational phenomenon?

(3) Are the terms or metaphors employed by Lorenz the most useful we can adopt?

(4) What are the implications for animal welfare of the assumptions built into the model?

In answer to the first question, there is indeed evidence that the tendency to perform certain behaviours increases as a function of the deprivation time. Lorenz gives several examples of this. What is the most appropriate language for describing these internal processes? Clearly, in the case of feeding and drinking, to a considerable extent one can anchor the internal change to physiologically realizable parameters, such as water level. However, that is not to say that, even for such systems, we can account for all of the deprivation effect in these terms (see later). The performance of the motor patterns associated with, say, feeding and drinking seems to have a component that is intrinsically of motivational value (reviewed by Toates, 1986).

Although one could argue whether 'energy' is the most appropriate or useful term to describe the internal factor, none-the-less the Lorenzian model captures the essential feature that intrinsic motivational changes can occur as a consequence of not performing a behaviour and that certain motor acts have their own intrinsic motivational potential. We might expect differences between motivational systems in this regard. For example, whereas evolutionary processes might have been able to leave most of the changes in drinking tendency that occur as a function of time to changes in hydrational level, it is more difficult to see what extraneural physiological parameter could underlie some other motivations. For example, the tendency of hens to engage in dustbathing increases as a function of time since they are deprived of the opportunity (Vestergaard, 1982).

Concerning the implications of the Lorenz model for animal welfare, one can identify 2 distinct areas. Firstly, the notion of 'energy accumulation' has negative connotations (Wiepkema, 1985a); one might logically assume that a failure to discharge the energy so as to confine it within bounds is associated with disturbance to the well-being of the animal. This might be so. Who knows? The problem is that we have no source of useful evidence either way. Whether an animal brought up in complete isolation suffers because of a failure to discharge a sex drive is perhaps more of a philosophical, than a biological, question. The second possible application of the Lorenzian assumptions is more subtle and perhaps more realistic. If the tendency to engage in behaviour of a particular kind increases as a function of deprivation, given even a minimum incentive cue or memory of one, we risk that the animal will perform, or attempt to perform, the behaviour in an abnormal way. This might cause damage. Even though we cannot show that any accumulation of metaphorical energy is intrinsically harmful the specific behavioural consequences of denying an animal the opportunity to perform the appropriate species-typical behaviour can sometimes be harmful. For example, hens maintained without litter attempt to dustbath in their feeding tray. To take another example, in the extreme, we might (wrongly) assume that the needs of the feeding system could be adequately maintained by supplying energy intravenously. As a somewhat less extreme case, we might invite a young animal to ingest nutrients taken from a bucket rather than by teat, assuming that the motor mechanisms of gaining food are a matter of indifference. Some harmful consequences of such procedures are described later.

Sensitization of the internal factor by deprivation might lead to excessive effort expended in pursuit of an unattainable goal. This might be stressful. In this context, it is relevant to mention the review of Mather (1981), in which she marshalls considerable evidence to show that so-called spontaneous activity in mammals (e.g. running-wheel activity in rats) is explicable in terms of an attempt to reach an unattainable goal. Such activity depends upon both internal physiological state and external cues concerning the location of incentives.

The discussion now turns to the current psychological literature where it is possible to see some models of the reinforcement process that offer hope of bridges with continental European ethology.

Reinforcement

Traditionally, in psychology the term 'reinforcement' has been applied to the consequences to an animal of its behaviour as far as the world external to the animal is concerned. For example, one might say that, to a food-deprived animal in a Skinner-box, food constitutes positive reinforcement. However, more recently, certain psychologists have come to recognize that behaviour *per se* can have reinforcement value (Herrnstein, 1977). Animals derive reinforcement from performing certain behaviours; in the other words, in the case of ingestive behaviour, reinforcement would derive in part from the chemical properties of the ingested substance and in part from the mechanics of ingestion. For example, in the case of a cat, the reinforcement of obtaining prey lies, in part, in stalking and hunting. The taste of mouse in the mouth is only a part of the complex of what constitutes reinforcement.

In a similar vein to Herrnstein, in their influential review, Glickman and Schiff (1967) wrote:

"The significance and magnitude of non-feeding chewing responses is not often appreciated by psychologists working with rats maintained in the usual barren environment (unless the rat happens to chew down the edges of their metal food dishes)".

The line taken by Herrnstein would seem to dovetail with some of the conclusions drawn by Kiley-Worthington (1977), regarding the welfare of domestic animals:

"One of the hazards of animals housed in controlled environments is that they are prevented from performing many of the normal activities which would occupy them during the day outside. Thus, their food is provided, they do not have to search for it or even graze for it, water is also provided. They do not have to take behavioural measures to keep themselves at a reasonable temperature. It is suggested that the provision of the methods for the animals to perform some of these tasks for themselves would help to keep the environmental stimulation level at its optimum. In addition, if suitably designed, such measures could cut down labour costs."

Homeostasis

A great deal of thinking in physiology, psychology and ethology is based upon the central principle of homeostasis. It is known that the important parameters of the body, e.g. water, sodium, temperature, cannot depart far from optimum. To do so, endangers the integrity of the organism. Furthermore, it is well known that deviation from optimum triggers intrinsic (autonomic) and behavioural action that serves to return the parameter to normal. Models based upon this negative-feedback principle have powerful predictive value (Cabanac and Russek, 1982). On the basis of the homeostatic model, one might conclude that deviations from optimum can be associated with negative affective states, and hence that welfare requires that the animal is able to maintain its parameters near to normal. Of course, in the extreme, tissue damage results from a failure to meet the homeostatic demands.

The homeostatic model, as traditionally described by analogy with negative feedback control systems in engineering, captures an essential feature of motivational systems that are associated with homeostasis. However, this particular analogy misses the peculiar mode by which the living system regulates. It does so by modulation of incentive properties. As evidence for such a model, consider the finding that rats infused intravenously with large amounts of water do not stop drinking (Nicolaidis and Rowland, 1975). Regulation of water and energy levels is achieved by a particular means, commerce with incentives and the use of specific motor patterns (e.g. chewing and licking). There is no reason to suppose that maintenance of needs by any other means is as satisfactory. Rats that learn to self-regulate their fluid intake by bar-pressing for intragastric reward are observed to lick and chew the lever at the time of bar-pressing (Altar and Carlisle, 1979). One could argue that (a) chewing and swallowing have their own motivational value and contribute to satiety, and (b) the animal is able to associate mock ingestive activities with nutrient gain (see earlier discussion of 'constraints on learning'). In these terms, satiety is a conjoint function of events at several levels in the system, the oral chemo-detection, gastric detection, post-absorptive detection and the mechanics involved in gaining the food. In rats, gastric emptying depends upon the animal's performance of the mechanics of chewing (Deutsch, 1979). Such results and the theoretical model that they support are potentially of crucial relevance to animal welfare. For example, in the case of veal calves, de Wilt (1985) notes:

"The performance of sucking during or immediately after milk intake may contribute to satiety and therefore sucking seems beneficial to calf welfare. If milk cannot be supplied through teats, adequate sucking facilities such as artificial teats should be presented in close proximity of the milk bucket. These teats are sucked intensively, which demonstrates their usefulness."

De Wilt notes that digestive processes are better stimulated by teat feeding than by bucket feeding. Careful consideration of the mode of food presentation, developmental stages and the animal's natural behaviour that serves food gain, enables the pathological sucking of other calves to be considerably reduced, if not eliminated. In normal feeding, an anticipatory secretion of hormones and digestive juices occurs prior to arrival of nutrients in the gut (see discussion in Rodin, 1981). It might be prudent to consider that we by-pass the foraging, chewing and swallowing components of the ingestive sequence at our peril.

Some have attempted to define stress simply in terms of deviations from normal of homeostatically regulated parameters (see discussion in Miller, 1980). Here some caution is in order. A fall in, say, body-energy level might well be to the animal's detriment, but we need to exercise caution in assuming it is a stressor *per se*, since deprivation is easily confounded with other factors. Thus, in an experiment on monkeys, Mason, Maher, Hartley, Mougey, Perlow and Jones (1976) showed greatly increased levels of urinary 17-OHCS during fasting. However, this was observed in colony-housed animals that were exposed to the cues normally associated with feeding. When they were housed under isolated conditions and given non-nutritive pellets at the time of anticipated feeding, increased 17-OHCS levels were not seen.

The importance of incentive presentation in the context of homeostatic regulation was underlined in an experiment on the repeated presentation of food/water following a period of food/water deprivation (Levine and Coover, 1976). In rats, following 24hr deprivation, plasma corticosterone level was greatly elevated. Returning the missing commodity caused a rapid fall in corticosteroid level. In fact, merely returning an empty water-spout caused a transient fall, followed by a sharp rise. The rapidity of these effects led to the assumption of a conditioned effect, since even when food and water were provided, the effect was too rapid to be accounted for in metabolic terms. The subsequent dramatic rise in corticosterone level seen by those animals only given a spout was attributed to a failure of the expectancy to be met. This experiment demonstrates the importance of avoiding frustrating and thwarting situations for animals.

Hedonic theories of motivation

In a series of experiments, Cabanac (1971, 1979 see also the chapter by Cabanac in this volume) showed the relationship between internal state and the reported affective values of such stimuli as heat and food. Humans were asked to rate hedonic value on a scale, and thus Cabanac was able to quantify affective state.

In accordance with everyday subjective experience, humans reported changes in affective rating, between positive, negative and neutral, as a function of internal state. The same food stimulus that might be perceived as hedonically positive following starvation would be said to be hedonically negative in surfeit. This line of investigation is relevant to us for several reasons. Although, of course, we cannot ask animals for verbal reports of their affective state, we can observe behavioural indices. Cabanac has shown a close correspondence between (a) human subjective reports and behavioural reactions and (b) the behavioural reactions of animals when confronted with incentives in various physiological states. In a more recent development of this work, Berridge and associates (Berridge and Grill, 1984) have produced a taxonomy of the rat's reaction to substances as a function of internal state. One is led to certain clear indices of positive and negative affective reactions.

The Opponent-process Theory of Motivation

The opponent-process theory of motivation proposed by Solomon and Corbit (1974) is relevant to the present discussion in the following way. Solomon and Corbit documented a large number of situations in which removal of a stimulus associated with one affective state sets up an affective state of the opposite sign. This model gives a formal scientific representation to what is a common-sense view: that just as stimuli can influence behaviour by their presentation, so can they by their withdrawal. For example, a chick that is imprinted on its mother and is then separated from her will emit distress calls and is capable of learning an instrumental task to regain contact. This might prove useful in understanding some features of welfare. For most mammalian species, enforced separation from the mother is a traumatic experience. First, a phase of 'protest' or 'agitation' is shown, characterized by, for example, increased vocalization. In some primate species, this phase is followed by a 'despair' phase of inactivity (see Kraemer, Ebert and McKinney, 1983). In rats, there is evidence that isolation from conspecifics causes stress, as indexed by ACTH secretion (discussed by Kiley-Worthington, 1977). As Dawkins (1980) notes, separating an individual sheep from the rest of the flock causes a large rise in the level of cortisone-type hormones.

Epstein's criteria of motivation

In an interesting review, Epstein (1982) draws a contrast between, on the one hand, the relative inflexibility of reflexes and instincts, and, on the other, the flexibility of motivational systems. At one extreme, reflexes are tied to particular stimulus inputs, the input is necessary for the behaviour to appear, and the behavioural form is stereotyped. By contrast, at the motivational systems end, we see (a) considerable autonomy from immediate stimulus control, (b) modulation by internal state and (c) the synthesis of novel means to gain access to the appropriate incentive object. For motivational systems, the animal's behaviour is guided by a goal and there exists an expectation of the outcome of the sequence of motivated behaviour.

Epstein argues that animals displaying motivation "behave with affect". He goes on "Motivated behaviour is I believe, often laden with affect, and affective expression should be expected of the internal state that is generating the behaviour." Epstein notes that (a) affect is shown by very young animals, (b) behaviour judged to be affectively loaded can sometimes be exhibited in its full form the first time that it appears and (c) is typically species-specific in so far as the outward expression of affective state is concerned. He indexes 'affect' by the presence of "discernable patterns of somatic and autonomic-glandular (both exocrine and endocrine) responding that are expressed as integral aspects of appetitive-consummatory sequences of behaviour."

Epstein associates the full complex of autonomic and behavioural expression that is labelled 'affective' with relatively advanced animals engaged in motivated activities. By contrast, he speculates that one would not see such a complex in, say, a cricket emitting its mating call. Epstein supposes that this is a "cooly mechanical instinctive behaviour." As particularly clear examples of affective display, Epstein gives those described by Darwin and Hess. In Darwin's work on the expression of emotions, cats are illustrated in clear poses of attack and soliciting. In response to electrical brain stimulation, the cats used by Hess showed unambiguous indices of internal state (e.g. rage or sleeping). To summarize his argument, Epstein writes: "....motivated behaviour is hedonic. It arises from mood, is performed with feeling, and results in pleasure or the escape from pain, and although the moods, feelings, and satisfaction themselves are private and beyond our reach as scientists, their overt expression in patterns of somatic, autonomic, and glandular responding is an important diagnostic characteristic of motivation."

The differences that Epstein supposes exist between instincts and motivational systems are shown in Table 1.

TABLE 1 Classification proposed by Epstein (1982)

Instinct	Motivation
Species-specific in releasing stimuli and efferent programs	Appetitive behaviour can be individuated
Appetitive behaviour unmodified by expectancy	Anticipatory of goals
Nonaffective	Accompanied by expression of affect
Present in a relatively large number of species	Present in relatively few species

Whether this represents a true dichotomy or rather two ranges on a continuum, is an interesting question, but one that need not concern us here (see Toates, 1986). The relevant point is that motivational systems share certain common properties, including affective states. Epstein draws a contrast between feeding in the fly, as an instinct, and the reaction of a duckling to an imprinted object, as an example of a motivational system. The behaviour of the fly is inflexible and stereotyped; it shows no evidence of learning in its food-getting behaviour. By contrast, in the case of the duckling, (a) the reaction of the bird to the object changes during imprinting, (b) the duckling can employ operants to regain contact with the imprinted object, (c) on separation, affect is shown, as indexed by distress calling.

Note that Epstein is not denying that some reflexes might have affective concomitants. Clearly, in the case of humans, removal of the foot from a thorn has both automatic reflexive aspects and an associated conscious pain perception. What Epstein is stressing is that motivational systems, as a feature of their organization, are affect loaded. Thus thwarting of the expression of goal-directed behaviour would be expected, in these terms, to be associated with negative affect.

Wiepkema's model

A model of behaviour described by Wiepkema (1985a; see also the chapter by Wiepkema in this volume) leads to some rather clear predictions regarding welfare and stress. The central assumption of Wiepkema's model is that the organism needs to be viewed as a special type of control system. It attempts to control both its external and internal environments to match certain internally specified values. In this model, in the case of a failure to control its Umwelt (the individual-specific or species-specific environment) in a way prescribed, the animal attempts, at least, to predict Umwelt changes. Wiepkema employs the German expression Sollwert ('should-value', desired-value, or set-point) to refer to the command to the system. The term Istwert refers to the actual state of the Umwelt at any point in time. The function of the system is to obtain equality between Sollwert and Istwert. Serious disparity would be associated with negative affect, if no means are available to eliminate the disparity. Equality or a move towards equality would be associated with positive affect.

What then is the evidence that animals set goals? A number of pieces of evidence point to the validity and usefulness of such a model (see also earlier discussion). Rats running mazes show evidence of having expectations regarding the goal object (Tolman, 1932). Signs of frustration, similar to those of fear are exhibited by an animal that arrives in a goal box to find the reward lacking or inadequate (for reviews, see Archer, 1976; Gray, 1975; Mackintosh, 1983; Toates, 1986). Such frustration is associated with a rise of plasma corticosterone levels, similar to that caused by a fear eliciting stimulus (Levine, Goldman and Coover, 1972). Changing the value of a goal object between sessions influences the animal's reaction when next placed in the situation (Dickinson, 1980; Tolman and Gleitman, 1949). An animal can employ a rich variety of different techniques (motor acts, strategies) in order to reach a given goal-object (Bindra, 1978; Toates, 1986). All of this leads to a picture of the animal setting goals, employing flexible strategies to attain those goals and experiencing negative affect on failing to attain them.

Powers (1971) is perhaps the best known exponent of the view that the brain of a vertebrate sets goals, and behaviour is controlled in such a way that sensory input matches, as far as possible, the goal ('input' or Sollwert) set. Such a model was given a neurophysiological embodiment by Cools (1985), where it was shown to fit rather well to the neurophysiological and behavioural evidence.

The notion of goal-setting is close to that of the 'self-initiation' of behaviour. This expression should not be taken to be a mystical one that places behaviour beyond the reach of scientific investigation. Rather, it should be seen as an affirmation of a model in which (a) behaviour is determined by internally specified set-points and (b) commerce with a given incentive might have very different effects, depending upon whether the animal initiated the commerce or was forced into it. This argument clearly needs some sharpening to make it scientifically respectable, but an example can rather neatly illustrate its essence.

Working with mice, Misslin and Cigrang (1986) obtained evidence to show that fear of novel objects is not caused by novelty *per se*, rather by the animal being *forced* into contact with novel objects. These workers looked at the physiological parameters of fear: defecation, urination and plasma corticosterone level. These indices did not show an increase in fear when the animal was free to initiate exploration of a novel area and could return to its home-base. However, placing the animal in the novel environment or impeding its return to the home-base did cause indices of fear to be exhibited. They conclude that:
"...fear or anxiety can be considered as a class of emotional reactions occuring when animals are unable to display normal behavioural sequences towards novel stimuli."

MODELS OF STRESS AND COPING THAT ARE BASED UPON LEARNING THEORIES

Prediction, control and feedback

Overmier, Patterson and Wielkiewicz (1980) note: "Control over one's destiny, knowledge of the "laws of nature", privilege of choice, and freedom from conflict are among the higher values and goals espoused by individuals. Are these values culturally determined and uniquely human, or are they rooted in the evolutionary biology of the organism?" The models described in this section start from the fundamental premise that animals are able to form predictive associations regarding events in the external world (Levine, Weinberg and Ursin, 1978; see also earlier discussion of contemporary animal learning theory). Levine *et al* (1978) relate their theoretical position to the work of Sokolov on habituation. In response to an unexpected stimulus a subject exhibits an alerting or orientation response, accompanied by an increase in the level of circulating adrenal corticoids. However, if the stimulus is not of a noxious quality and is repeated a number of times, the signs of activation diminish. The system is said to be habituated. In this model the organism matches incoming sensory information against internal models or representations of stimuli. These give expectancies of what is arising in the environment.

In response to aversive stimuli, the signs of stress are greater if the aversive stimulus is unpredictable as compared to predictable (reviewed for a variety of species by Broom, 1985). However, Broom's comment that "Unpredictable input or the absence of expected input means that the animal is not regulating its life adequately and hence such unpredictability can be disturbing" needs a slight qualification. Omission of an expected aversive event creates a state that contemporary animal learning theorists describe as 'relief'. It has several of the same properties as presentation of a positive incentive object, and there is little or no reason to think it would be aversive.

As a rule, rats and dogs show a clear preference for prediction and control as compared to unpredictability and lack of control (Overmier, Patterson and Wielkiewitz, 1980). Rats will work for food in the presence of free food, and show a preference for signalled shock over unsignalled shock. A shock that is predictable causes the development of fewer ulcers in rats, than does unpredictable shock (Gliner, 1972). As Levine, Weinberg and Ursin (1978) note, what is apparently the same physical stimulus can have different effects depending upon whether it is predictable. For some time, there were two hypotheses that attempted to explain an animal's preference for signalled shock (see discussion in Levine, Weinberg and Ursin, 1978). One notion is that information is provided to the subject by the warning signal such that some kind of response that reduces the impact of the shock is possible. This is known as the _preparatory response hypothesis_. A different, but not necessarily mutually exclusive hypothesis, is termed the _safety-signal hypothesis_ or _safe-unsafe_ hypothesis. The basis of this hypothesis is the assumption that the predictive signal enables the animal to identify shock-free periods. During such periods, relaxation is possible. In terms of contemporary learning theory (Dickinson, 1980), the animal would learn an association of the form 'absence of a signal predicts no shock'. More recently, another factor has entered the picture. Fanselow (1979) found that the rat's preference for signalled shock is blocked by administration of the opiate antagonist, naloxone. This suggested to him that, by a process of Pavlovian conditioning, the warning cue came to elicit a release of opiates that attenuated the negative affective value of the shock.

Exposure to the presentation of biological incentives (e.g. positive ones such as food and negative ones such as shock) that is uncontrolled makes later learning of a contingency in which the animal does have control more difficult. This interference generalizes between positive and negative incentives, which suggests that a general mechanism of behavioural control has in some way been disrupted (Overmier et al, 1980). Later theories by Overmier and associates proposed that there are three consequences of exposure to uncontrollable, inescapable, noxious stimuli: (a) the animal's emotional state is altered, there being a tendency towards emotional depression and generalized anxiety, (b) there is a decreased motivation to respond in the face of aversive stimuli and (c) there is a deficiency in subsequent learning of response-reinforcer contingencies.

In an experiment by Weiss (1971), it was shown that rats that were able to perform a coping response to postpone, avoid or escape shock were less stressed (by an index of gastric ulceration) than matched controls subjected to the same number of shocks but unable to influence them. This result leads us to reiterate a point made earlier that the stress induced by a stressor depends upon the whole motivational context of its presentation.

One of the most stressful events, as indexed by plasma corticosterone, is to remove control from an animal that had previously experienced control over its situation (reviewed by Overmier, Patterson and Wielkiewicz, 1980). This is more stressful than exposure to the same aversive conditions of a group of animals never having experienced control. Such loss of control is created by, for example, presenting unavoidable shocks to animals after their learning of an avoidance response. Even removing the manipulandum with which an animal has learned to exert control, but delivering no further shocks, causes greatly elevated steroid levels.

Blocking the performance of shuttlebox avoidance by a barrier and the imposition of extinction conditions in an appetitively motivated task both increase plasma corticoid levels. Weinberg and Levine (1980) suggest that both situations can be subsumed under the heading of 'loss of control'.

Feedback informing the animal that by means of the contingency in operation it is now in a safe period is an important factor in diminishing stress (see Weinberg and Levine, 1980). Such feedback might be response dependent, e.g. a sound that is unambiguously associated with safety and which is emitted when the animal presses a bar that terminates shock.

Ursin's model

Ursin (1985) presented a model of emotional behaviour that is very relevant to the present argument. In this section, I shall describe in some detail this model, the assumptions on which it is based, and its predictions.

The first assumption Ursin uses is one articulated by Bolles (1972). Bolles proposes that learning is of basically two kinds; animals learn the relationship between either (a) two events or (b) their own behaviour and the consequences of this behaviour (see also earlier discussion). These are termed 'stimulus expectancies' and 'response expectancies', respectively. In such terms, Ursin assumes that "brains are able to store the information that certain stimuli or responses precede other stimuli." For example, a rat in a Skinner box learns that the response of bar-pressing causes the arrival of food. Ursin's second assumption regards the process of activation. Ursin defines activation as a "general, nonspecific response to all situations in which there is a threat to the organism". Correlated with changes in activation level are the blood concentrations of many hormones, e.g. cortisol. He notes that activation can also appear in situations where expectancies are not met. According to Ursin:
"Activation occurs in all situations where the brain registers a discrepancy between a set value for one of the many variables controlled by the brain." He also refers to 'emotional behaviour' being aroused in the same situations as cause activation. As an index of activation, according to Ursin, one can use the electroencephalogram (EEG). He refers to activation as occurring whenever "there is something wrong". According to his formulation, when there is a discrepancy between set value and actual value, activation will normally continue until the discrepancy is eliminated. This would normally be by means of driving the actual value into equality with the set value. Under some circumstances it might be achieved by adjusting the set value. Ursin defines instrumental behaviour as that which "reduces discrepancies between set values and actual values for variables controlled by the brain."

Ursin notes that a hungry rat, put into a small cage, will roam around for a while. When the rat has established that there is no food to be found, it will stop searching and its corticosterone level will drop. Signs of activation will, however, be revived by a change in stimulus situation that could herald the arrival of food.

Ursin suggests that there might exist an optimal level of activation, such that levels that are too low might also cause action to be instigated to increase the level.

A number of terms and theoretical concepts arise in the discussion of Ursin's model. The animal is said to be coping when it can perform a response that will eliminate a threat. Thus activation is increased in a threatening situation, and performance of a coping response decreases activation. In a well-trained sequence involving predictable events, having a coping response can largely preempt activation. In rats, the ability to perform fighting in response to shock is associated with lower corticosterone levels than controls unable to perform a response. In an avoidance learning task, rats at first show a rise in corticosterone levels over sessions of exposure, but as they become accomplished at the task there is some decline seen (Coover, Ursin and Levine, 1973). Blocking the ability to perform the avoidance response, by a partition in the shuttle-box, and presenting the warning cue but no shock is associated with an increase in corticosterone level. Dominant ptarmigans show lower corticosterone levels following an agonistic encounter than do more submissive individuals. Ursin suggests that a history of winning fights is associated with a greater response-outcome expectancy.

In terms of behaviourist motivation models, a drop in activation level, that serves to reinforce a successful coping response would be termed 'drive reduction' (Ursin, 1980; Miller, 1980).

The definition of coping adopted by Levine, Weinberg and Ursin (1978) is "based on the ultimate reduction of the physiological arousal produced as a consequence of the novelty or threat of any given stimulus complex." Such coping would normally be associated with a behavioural response that puts a distance between the animal and the threat. However, if no such response is observed, and yet by the physiological index the animal is not in the aroused state, Levine et al. would still term this 'coping'. Ursin (1985) defines the term hopelessness as describing the situation where, irrespective of the behaviour of the subject, an aversive event will follow. It is the opposite of coping. The term 'helplessness' refers to the situation where there is no relationship between a response and the outcome.

BEHAVIOURAL PHENOMENA AND THE RELEVANCE OF MODELS

Conflict

Stress, as indexed by gastric ulceration, is increased if the animal is placed in an approach-avoidance conflict (Sawrey, Conger and Turrell, 1956; see also review in Murison, 1980). For example, rats that can only get to food by passing over an electrified grid are in such an approach-avoidance conflict. In principle, the animal has control over such a situation. However, the control is bought at a price: to eat it must pay the price of passing an aversive region, and to be free of shock it must pay the price of hunger.

Stereotypies

The frequency of stereotypies in confined sows is greatly reduced by injections of the opiate antagonist naloxone (Cronin, Wiepkema and van Ree, 1986; see also discussion in Broom, 1985). One possible interpretation of this result is that animals are able to increase analgesic levels by the performance of stereotypies. Presumably the housing conditions are made less aversive in this way. At this point, observations on domestic animals would appear to dovetail with some findings on opiate taking in rats. Rats isolated in small cages were found to drink 16 times more morphine than rats housed in a colony (Alexander and Hadaway, 1982). These authors noted:

"The restrictive, isolated conditions of standard laboratory housing may be inherently stressful to mobile social animals like rats..."

Opiates would therefore be a vehicle for attaining a more acceptable environment or 'affective tone', depending upon ones viewpoint.

In poultry, thwarting access to food is well recognized as a cause of aggression and stereotypies (see review in Broom, 1985). Pigs housed alone and confronted with a situation of thwarting (e.g. prevention of access to food or non-appearance of expected food) attempt to escape from the situation. Such situations of frustration are commonly encountered in farms. Broom (1983) discusses the possible mechanism underlying, and function of, stereotypies. He notes that stereotypies cause a change in sensory input. This might serve to maintain arousal level. Furthermore, since the animal is able to generate a predictable sensory input by performing stereotypies, an unpredictable environment could be made more predictable.

"If the number of operational attention channels were limited, the individual might then reduce the necessity to process and respond to inputs from unpleasant stimuli."

Stereotypies are commonly exhibited by animals in either monotonous environments or in environments of extreme novelty or containing stimuli thought to be anxiety-inducing. Examples in the latter category include the presence of a potential predator, a delay in the time of an anticipated meal or removal of a social companion. In tethered sows, the stereotypies of chewing and licking parts of the cage can be reduced by making straw available. Broom uses such evidence to support the notion that stereotypies are a response to restricted and monotonous surroundings. Straw permits a normal activity, chewing, to occur. In other situations, the lack of a specific stimulus seems to evoke stereotypies. Broom offers as an example here pacing behaviour by hens denied access to a nest site and seen in the period just before laying. Broom concludes that "the occurrence of prolonged stereotypies indicates that the conditions are adverse for that individual". Of course, the adverse conditions might well become less adverse by the performance of the stereotypy. As Kiley-Worthington (1977) notes:

"For practical applications it must, therefore, be clearly appreciated that to change the environmental conditions to cure an established stereotypy may well have little effect. However, the provision of an 'adequate' environment of sufficient complexity to avoid insufficient sensory input may prevent its development.

Thus it is suggested that the performance of stereotypies should be regarded as a symptom of environmental inadequacy."

Environmental design

Personal space is an important factor in stress reduction. Overcrowding causes signs of stress in various species, and increases the probability of aggressive encounters (Kiley-Worthington, 1977). Welfare can sometimes be improved by provision of spaces where animals can hide from more aggressive conspecifics. A subordinate animal might get pushed out in the competition for food. In a natural environment, the response to overcrowding would be migration from the environment in question. According to Kiley-Worthington (1977), in farm animals, a common form of thwarting that provokes aggression is due to the distribution and type of food that is made available. Trough feeding can increase competition pressure. Dramatic changes in the social environment, e.g. isolation or exposure to unfamilar animals in a confined space, increases signs of stress in domestic animals (Kiley-Worthington, 1977). Kiley-Worthington proposes that for dairy cattle "groups should be as stable as possible, preferably being composed of the same animals from birth to death". Introductions between two animals should be carried out in a large area rather than a confined space. Stress and physical injury can often be avoided by making the introduction on mutually unknown territory (Kiley-Worthington, 1977). Whereas sudden stimuli (e.g. new personnel, loud noises) usually need to be avoided in order to minimize fear reactions, a monotonous environment is also to be avoided (Kiley-Worthington, 1977). A compromise would consist of a mild level of opportunity for changing stimulation. Sudden unexpected noises appear to induce stress, and for this reason regular background noises might be a desirable feature of an environment. The opportunity for self-determined environmental change could well improve welfare (e.g. the chance for cattle to stand in the rain, or for pigs to look in a mirror or for animals to play with objects and gadgets) (Kiley-Worthington, 1977).

DISCUSSION

After reviewing the experimental evidence and various models of motivation, it is possible to form some consensus of opinion. With the inroads made by cybernetics, cognitive psychology and cognitive ethology, particularly in the last 20 years, one can draw a portrait of an advanced vertebrate and make some conclusions regarding welfare.

The model of the rat as a stimulus-response (S-R) automaton has been largely replaced by that of a goal-seeking system. The term 'Sollwert' is often used to refer to the goals that the animal sets. The animal is able to take action via a variety of different routes in order to attain equality between Sollwert and Istwert. By analogy and homology with the case of humans, it would seem reasonable to associate negative affective states with a failure to attain a goal or a move away from a goal. Conversely, positive affective states would be associated with movements towards a goal. The animal is caused to move in a gradient, towards the attainment of positive affective states and the avoidance of negative ones.

The stimulus-response model might logically lead us to suppose that the animal is quiescent until a stimulus impinges upon it. We now believe that animals seek a certain minimum level of stimulation, and that under-stimulation can be aversive, just as over-stimulation can.

The notion of 'set-point', goal or Sollwert is, in some respects and situations, similar to that of expectation (Archer, 1976). However, it is not identical. Sollwert carries the notion of what the animal 'wants' or 'commands' from the situation - the state that is the logical outcome of its activity. Presentation of a cue predictive of shock might well produce an expectation of shock, but that would, of course, not be the Sollwert of the system. Thus one needs to qualify Archer's model of aggression. Omission of an expected aversive event is not a cue for aggression; indeed, relief is used to describe the state accompanying this event (Mackintosh, 1983). A very similar point to this was made by Goldman, Coover and Levine (1973) in reference to the elation induced by a surprising increase in the size of reward given to rats in an operant task and the associated drop in pituitary-adrenal level.

Contemporary animal learning theory dovetails with the model of goal-seeking. Animals are now seen to learn cognitions of the kind 'x is followed by y' and 'response x leads to event y'. This gives behaviour a flexibility that would be impossible with the S-R model. Animals exhibit a preference for situations in which they can predict what is happening in their environment, and this predictive ability is in accordance with the models currently popular in learning theory. The learning of successful escape or avoidance responses when confronted with an aversive situation or a cue predictive thereof is aided by feedback informing the animal that safety has been attained (Bolles, 1975). Similarly, stress seems to be reduced by presentation of such a cue (Miller, 1980).

Suppose we adopt a goal-seeking model. Can we obtain any theoretical synthesis and parsimony with such a model? Can we see common features between a number of different experimental situations? From the foregoing evidence, it would seem that in several different situations a stress reaction is caused by a failure to meet a goal or expectation.

(a) non-reward or a reward of value less than expected causes this reaction.

(b) somewhat similar to (a) is the blocking of access to, say, food, by the presence of a more dominant animal

(c) in a conflict situation, the animal would be expected temporally to set one goal, say, getting to water, only to find this in conflict with a different goal, say, avoiding shock.

(d) physical restraint induces stress (Thorpe, 1969), and it was suggested by Ader (1971) that the degree of stress depends upon the extent to which natural activity patterns (e.g. movement at estrus) are thwarted.

If a goal can be realized, then despite the fact that the animal is in a potentially dangerous situation, there is relatively little sign of fear or stress. For example, after an animal has learned an avoidance task it performs this in a calm manner (Bolles, 1975).

Traditionally, the homeostatic model has carried much explanatory weight regarding welfare. Indeed, one would logically associate suffering with deviations from the homeostatic norm that exceed a certain threshold. Some have defined stress in terms of deviations from homeostasis (see discussion in Miller, 1980). However, some caution is in order, and perhaps it is more useful to refer to 'stress' in terms of failure to meet goals as related to the external world. Seen in these terms, homeostatic mechanisms (e.g. energy regulation) set up goals (e.g. attain food), whose realization would be associated with positive affective states and whose thwarting would give rise to stress and associated negative affective states. For example, food deprivation would make the attainment of food a top-priority goal and competition for food would induce stress. In terms of contemporary theory, perhaps we can define stress as a state induced by a failure to have a response available for attaining a positive goal and/or avoiding a negative state.

From a variety of sources of evidence, we come to the conclusion that to have control over its environment is a fundamental need of a sophisticated organism. For example, rats will sometimes work for food in the presence of free food. In addition to having control as a property in its own right, this ability seems all the stronger, if by exploiting a mechanism of control, the animal can also avoid a negative event and/or attain a positive one. Perhaps we can define suffering in terms of lack of control: opening the feedback loop. We cannot assume that all activation of the system for coping with stress is aversive. Provided that the animal has a control strategy, involving a response-outcome expectancy that is not too demanding in terms of effort (Weiss, 1971), we ought not to be too concerned for the animal's welfare.

In its commerce with the external world, e.g. escape from noxious events and gaining of food or warmth, an animal tends to show species-specific patterns of behaviour. We should not under-estimate the importance of such behaviour; energy regulation is more than gaining calories; it involves a sequence of activities, such as capture of food, chewing and swallowing. The evidence suggests that such activities have their own intrinsic motivational potential, and are important for the whole regulatory system; recognition of this can provide a bridge between Lorenzian and cognitive models of behaviour.

By looking at the bases of motivation and affect at the level of neurophysiology and pharmacology, we might gain some insight into affective value, motivation and stress. In this context, we noted that (a) rats in a dull environment will work for opiates, (b) the opiate antagonist naloxone abolishes the rat's preference for signalled shock over unsignalled shock and (c) in confined sows the frequency of stereotypies is reduced by naloxone injections. Such results encourage the kind of model that sees the animal as being moved to maximize stimulation in certain affective circuits, in which opiates can play a major role.

With the assistance of the foregoing discussion of experimental evidence and theoretical models, one can start to formulate some practical suggestions regarding welfare. At present these are less concrete than might be desired, but they do appear to fit with practical experience. The suggestions are as follows:

(1) as far as is realistic, animals should be allowed to set their own goals and given the chance to realize these goals. For example, free exploration of a new territory is to be preferred to forced contact (However, this does not mean that animals always make the choice that best serves their survival. In the area of ingestive behaviour, some counter examples can be given. See discussion in Toates, 1986).

(2) whereever possible, it would be wise to allow an animal to exploit its own species-specific, and developmental-stage-specific, mode of commerce with the world. For example, animals that naturally gain food by rooting should be given the opportunity to do so. A young animal should not be denied the expression of suckling.

(3) animals should be given the maximum opportunity to exert control over their environments. Changes that the animal can initiate, such as changing the illumination level or gaining food by operant means, are to be encouraged.

Acknowledgement

I am grateful to Drs. Marion Dawkins and Paul Verrell for their helpful comments made during the development of the ideas expressed here.

REFERENCES

Ader, R. 1971. Experimentally induced gastric lesions. I. Experimental investigations. Advances in Psychosomatic Medicine, 6, 1-39.

Alexander,B.K. and Hadaway,P.F. 1982. Opiate addiction: The case for an adaptive orientation. Psychological Bulletin, 92, 367-381.

Altar, A. and Carlisle, H.J. 1979. Intragastric drinking in the rat: Evidence for a role of oropharyngeal stimulation. Physiology and Behaviour, 22, 1221-1225.

Archer, J. 1976. The organization of aggression and fear in vertebrates. In "Perspectives in Ethology" Vol.2 (Eds. P.P.G. Bateson and P.H. Klopfer). (Plenum Press, New York). pp.231-298.

Archer,J. 1979. Animals under Stress. (Edward Arnold, London).

Archer, J. and Birke, L.I.A. 1983. Exploration in Animals and Humans. (Van Nostrand Reinhold, Wokingham).

Berridge, K.C. and Grill, H.J. 1984. Isohedonic tastes support a two-dimensional hypothesis of palatability. Appetite, 5, 221-231.

Bindra, D. 1978. How adaptive behaviour is produced: a perceptual-motivational alternative to response-reinforcement. The Behavioural and Brain Sciences, 1, 41-91.

Bolles, R.C. 1972. Reinforcement, expectancy and learning. Psychological Review, 79, 394-409.

Bolles, R.C. 1975. Theory of Motivation. (Harper and Row, New York).
Bolles, R.C. 1979. Learning Theory. (Holt, Rinehart and Winston, New York).
Broom,D.M. 1983. Stereotypies as animal welfare indicators. In "Indicators Relevant to Farm Animal Welfare" (Ed. D.Smidt). (Martinus Nijhoff, The Hague). pp.81-87.
Broom,D.M. 1985. Stress, welfare and the state of equilibrium. In "Proceedings of the Second European Symposium on Poultry Welfare" (Ed. R.M.Wegner). (World Poultry Science Association, Celle). pp.72-81.
Cabanac, M. 1971. Physiological role of pleasure. Science, 173, 1103-1107.
Cabanac, M. 1979. Sensory pleasure. The Quarterly Review of Biology, 54, 1-29.
Cabanac, M. and Russek, M. Régulation et Contrôle en Biologie. (Les Presses de L'Université Laval, Quebec).
Cools,A.R. 1985. Brain and behaviour:Hierarchy of feedback systems and control of input. In "Perspectives in Ethology" Vol.6 (Eds. P.P.G. Bateson and P.H. Klopfer). (Plenum Press, New York). pp.109-168.
Coover,G.D., Ursin,H. and Levine,S. (1973) Plasma-corticosterone levels during active-avoidance learning in rats. Journal of Comparative and Physiological Psychology, 82, 170-174.
Cronin,G.M. 1985. The development and significance of abnormal stereotyped behaviours in tethered sows. Doctoral thesis, Agricultural University of Wageningen.
Cronin,G.M., Wiepkema,P.R. and van Ree,J.M. 1986. Endorphins implicated in stereotypies of tethered sows. Experientia, 42, 198-199
Dawkins,M.S. 1980. Animal Suffering. (Chapman and Hall, London).
Deutsch, J.A. 1979. Intragastric infusion and pressure. The Behavioural and Brain Sciences, 1, 105.
Dickinson,A. 1980. Contemporary Animal Learning Theory. Cambridge University Press.
Epstein,A.N. 1982. Instinct and motivation as explanations for complex behaviour. In "The Physiological Mechanisms of Motivation" (Ed. D.W. Pfaff). (Springer-Verlag, Berlin). pp. 25-58.
Fanselow, M.S. 1979. Naloxone attenuates rat's preference for signalled shock. Physiological Psychology, 7, 70-74.
Glickman,S.E. and Schiff,B.B. 1967. A biological theory of reinforcement Psychological Review, 74, 81-109.
Gliner,J.A. 1972. Predictable vs. unpredictable shock: preference behaviour and stomach ulceration. Physiology and Behaviour, 9, 693-698.
Goldman,L., Coover,G.D. and Levine,S. 1973. Bidirectional effects of reinforcement shifts on pituitary adrenal activity. Physiology and Behaviour, 10, 209-214.
Gray, J.A. 1975. Elements of a Two-Process Theory of Learning. (Academic Press, London).
Hellhammer,D. 1983. Learned helplessness - An animal model. In "The Origins of Depression: Current Concepts and Approaches" (Ed. J.Angst). (Springer-Verlag, Berlin). pp.147-161.
Herrnstein,R.J. 1977. The evolution of behaviourism. American Psychologist, 32, 593-603.
Hull, C.L. 1952. A Behaviour System. (Yale University Press, New Haven).
Jaynes, J. 1982. A two-tiered theory of emotions: affect and feeling. The Behavioural and Brain Sciences, 5, 434-435.
Katz,R.J. 1983. Stress, conflict and depression. In "The Origins of Depression: Current Concepts and Approaches" (Ed. J.Angst). (Springer-Verlag, Berlin). pp. 121-132.

Kiley-Worthington,M. 1977. Behavioural Problems of Farm Animals. (Oriel Press, Stocksfield).
Kraemer,G.W., Ebert,M.H. and McKinney,W.T. (1983) Separation models and depression. In "The Origins of Depression: Current Concepts and Approaches" (Ed. J.Angst). (Springer-Verlag, Berlin). pp.133-145.
Levine,S. and Coover,G.D. 1976. Environmental control of suppression of the pituitary-adrenal system. Physiology and Behaviour, 17, 35-37.
Levine, S., Goldman, L. and Coover, G.D. 1972. Expectancy and the pituitary adrenal system. In "Physiology, Emotion and Psychosomatic Illness" CIBA Foundation Symposium, 8 (Elsevier, Amsterdam). pp.281-296.
Levine,S., Weinberg,J. and Ursin,H. 1978. Definition of the coping process and statement of the problem. In "Psychobiology of Stress - A Study of Coping Men" (Eds. H.Ursin, E.Baade and S.Levine). (Academic Press, New York). pp.3-21.
Lorenz, K. 1966. On Aggression. (Methuen, London).
Lorenz, K. 1981. The Foundations of Ethology. (Springer, New York).
Mackintosh, N.J. 1983. Conditioning and Associative Learning. (Clarendon Press, Oxford).
Mather, J.G. 1981. Wheel-running activity: a new interpretation. Mammal Review, 11, 41-51.
Mason,J.W., Maher,J.T., Hartley,L.H., Mougey,E.H., Perlow,M.J. and Jones, G.J. 1976. Selectivity of corticosteroid and catecholamine responses to various natural stimuli. In "Psychopathology of Human Adaptation" (Ed. G.Serban). (Plenum Press, New York). pp.147-171.
Menzel, E. 1978. Cognitive mapping in chimpanzees. In "Cognitive Processes in Animal Behaviour" (Eds. S.H. Hulse, H. Fowler and W.K. Honig). (Lawrence Erlbaum, Hillsdale). pp.375-422.
Miller,N.E. (1980) A perspective on the effects of stress and coping on disease and health. In "Coping and Health" (Eds. S.Levine and H. Ursin). (Plenum Press, New York). pp.323-353.
Misslin,R. and Cigrang,M. 1986. Does neophobia necessarily imply fear or anxiety? Behavioral Processes, 12, 45-50.
Murison,R. 1980. Experimentally induced gastric ulceration: A model disorder for psychosomatic research. In "Coping and Health" (Eds. S. Levine and H.Ursin). (Plenum Press, New York). pp.281-322.
Nicolaidis, S. and Rowland, N. 1975. Systemic versus oral and gastrointestinal metering of fluid intake. In "Control Mechanisms of Drinking" (Ed. G. Peters, J.T. Fitzsimons and L. Peters-Haefeli). (Springer-Verlag, Berlin). pp. 14-21.
O'Keefe, J. and Nadel, L. 1978. The Hippocampus as a Cognitive Map. (Clarendon Press, Oxford).
Overmier,J.B., Patterson,J. and Wielkiewicz,R.M. 1980. Environmental contingencies as sources of stress in animals. In "Coping and Health" (Eds. S.Levine and H.Ursin). (Plenum Press, New York). pp.1-38
Panksepp, J. 1982. Toward a general psychobiological theory of emotions. The Behavioural and Brain Sciences, 5, 407-467.
Powers,W.T. 1971. A feedback model for behaviour: Application to a rat experiment. Behavioural Science, 16, 558-563.
Rodin, J. 1981. Current status of the internal-external hypothesis for obesity. American Psychologist, 36, 361-372.
Sawrey,W.L., Conger,J.J. and Turrell,E.S. 1956. An experimental investigation of the role of psychological factors in the production of gastric ulcers in rats. Journal of Comparative and Physiological Psychology, 49, 457-461.
Solomon, R.L. and Corbit, J.D. 1974. An opponent-process theory of motivation. I. Temporal dynamics of affect. Psychological Review,

81, 119-145.
Thorpe, W.H. 1969. Welfare of domestic animals. Nature, 224, 18-20.
Toates, F.M. 1983. Exploration as a motivational and learning system: A cognitive incentive view. In "Exploration in Animals and Humans" (Eds. J. Archer and L.I.A. Birke). (Van Nostrand Reinhold, Wokingham). pp. 55-71.
Toates,F.M. 1986. Motivational Systems. Cambridge University Press.
Tolman, E.C. 1932. Purposive behaviour in animals and men. (The Century Company, New York).
Tolman,E.C. and Gleitman,H. 1949. Studies in learning and motivation: I. equal reinforcements in both end-boxes, followed by shock in one end-box. Journal of Experimental Psychology, 39, 810-819.
Ursin,H. 1980. Affective and instrumental aspects of fear and aggression. In "Functional States of the Brain and their Determinants" (Eds. M.Koukkou, D.Lehmann and J.Angst). (Elsevier, Amsterdam). pp.119-130.
Ursin,H. 1985. The instrumental effects of emotional behaviour. In "Perspectives in Ethology" Vol.6 (Eds. P.P.G. Bateson and P.H. Klopfer). (Plenum Press, New York). pp. 45-62.
Vestergaard,K. 1982. Dust-bathing in the domestic fowl - Diurnal rhythm and dust deprivation. Applied Animal Ethology, 8, 487-495.
Walker, S. 1983. Animal Thought. (Routledge and Kegan Paul, London).
Weinberg,J. and Levine,S. 1980. Psychobiology of coping in animals: The effects of predictability. In "Coping and Health" (Eds. S.Levine and H.Ursin). (Plenum Press, New York). pp.39-59.
Weiss,J.M. 1971. effects of coping behaviour in different warning signal conditions on stress pathology in rats. Journal of Comparative and Physiological Psychology, 77, 1-13.
Wiepkema, P.R. 1985a. Abnormal behaviours in farm animals: Ethological implications. Netherlands Journal of Zoology, 35, 279-299.
Wiepkema, P.R. 1985b. Biology of fear. Proceedings of the Second European Symposium on Poultry Welfare (Ed. R.M.Wegner). (World Poultry Science Association, Celle), pp.84-93.
Wilt,J.G. de 1985. Behaviour and welfare of veal calves in relation to husbandry systems. Institute of Agricultural Engineering, P.O.Box 43, 6700 AA Wageningen, Netherlands.
Young, P.T. 1966. Motivation and Emotion. (Wiley, New York).

THE PLACE OF BEHAVIOUR IN PHYSIOLOGY

M. Cabanac

Université Claude Bernard
Faculté de Médecine Lyon-Sud
Laboratoire de Physiologie
B.P. 12 69921 Oullins Cedex, France

ABSTRACT

From the long-term point of view the flow of energy and matter between animals and their environment is behaviourally controlled either as inflow or outflow. It is argued here that such adaptation of behaviour to physiology is obtained by the tendency to maximize sensory pleasure.

There is a permanent exchange-flow of matter and energy between living beings and their environment. Without this exchange, it is not possible to understand physiology. Inflow and outflow may be controlled by purely autonomic functions such, for example, as pulmonary ventilation. However, most of this exchange is controlled by behaviour. Food and water intake, defecation and micturition are behaviourally controlled. Heat loss and to a lesser extent heat gain are also behaviourally controlled. Finally all external work, whatever its purpose, is also a behaviour.

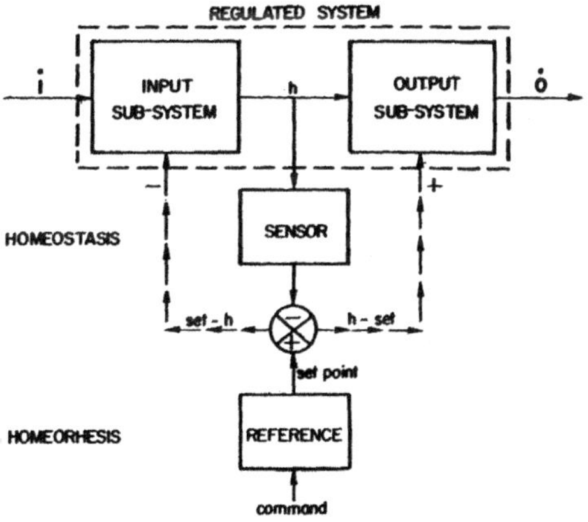

Fig. 1 Block diagram analysis of the place of behaviour in physiology (From Cabanac & Russek, 1982). Any regulated variable h depends on inflow and outflow subsystems. Behaviour (interrupted arrows) acts on inflow with a negative feed-back and on outflow with a positive feed forward. Regulatory behaviour is triggered by the difference between regulated variable h and set point. Homeorhesis (Nicolaïdis, 1977) is the long-term adjustment of the set point (e.g. body weight annual cycle).

Figure one illustrates this behavioural control. In addition, this diagram emphazises that control of inflow and outflow are two separate processes. An organism gains thus one degree of freedom by adjusting inflow when outflow is imposed and reciprocally.

One of the main questions put to the physiologist by the observation that behaviour is the necessary first step and last step of whole-body physiology, is how such finality is met (Toates, 1986). Experiments on humans cast light upon this process.

When a stimulus excites a sensory neuron it may arouse a sensation. This sensation is tri-dimensional. The first dimension of sensation is qualitative, identifying the nature of the stimulus. The second is quantitative, describing the intensity of the stimulus. The third, which may be absent, is affective. The affective part of a sensation is the amount of pleasure or displeasure aroused by the stimulus. According to Young (1959), this dimension of sensation is a continuum between two extremes : extreme negative affectivity, or distress, and extreme positive affectivity, or delight. Midway between the two lies indifference.

Not all stimuli evoke pleasure or displeasure. In the vast, permanent flux of inputs from the sensors to the central nervous system, the large majority elicits an indifferent sensation. For example, the sight of most objects is neither pleasurable nor displeasurable, as such. If affectivity is involved, the source is a sense of esthetics. In addition, as stated by Pfaffmann (1960), "there is almost no limit to the range of previously neutral stimuli that, by one method or another, can be made pleasurable or unpleasurable". Pfaffmann designated as primary reinforcers all stimuli creating sensations of pleasure or displeasure. The majority of these are negative reinforcers. A stimulus can be unpleasant by its very nature (e.g. a bitter taste) or by its intensity (e.g. a violent sound). When the intensity of a neutral stimulus increases, the sensation usually becomes unpleasant, as was suggested by Wundt (1874). Thus, it can be stated that pleasant stimuli are on the minority; it is striking to observe that the range of pleasantness is limited both qualitatively and quantitatively. The most effective, and probably only, primary positive reinforcers are chemical, thermal, and mechanical stimuli.

There is an obvious relation between the affective part of sensation and behaviour ; the strength of the motivation for or against a stimulus is a function of the intensity of the pleasure or displeasure elicited by the stimulus. Pleasant sensations induce approach or "consumption" behaviours (or both) for alimentary, sexual, and thermal stimuli. Relations exist between pleasure and usefulness, and between displeasure and harm or danger. The evidence supports the hypothesis that sensory pleasure is a sign of usefulness and displeasure a warning sign. In the case of thermal stimuli, the pleasure or displeasure aroused by a stimulus can be predicted, from the various body temperatures of the person stimulated. Pleasure is actually observable only in transition, when the stimulus aids the subject to return to normothermia. As soon as the subject returns to normothermia, all thermal stimuli lose their strong pleasure component and tend to be indifferent or unpleasant. This scarcity of pleasure may be more apparent than real, because temperature regulation is never achieved, and normothermia is an almost virtual situation.

Alliesthesia : Coined from the Greek roots (Allios- changed and -esthesia sensation), this word is applied to the affective component of sensation, pleasure or displeasure (Cabanac, 1971). The amount of pleasure or displeasure aroused by a given stimulus is not invariable, but depends on the internal state of the stimulated subject. Factors that can modify the internal state and, in turn, induce alliesthesia are (i) internal physiological variables, e.g. deep body temperature (Attia, 1984) or body dehydration (Rolls et al. 1980) modifying the pleasure of thermal sensation or taste of water ; (ii) set points, e.g. during fever the body temperature set point is raised and pleasure defends the elevated set point ; (iii) multiple peripheral stimuli (Rolls, 1981 ; Marks and Gonzalez, 1974), e.g. mean skin temperature determines the set point for deep body temperature and generates alliesthesia ; (iv) past history of the subject (Garcia, Hankin and Rusiniak, 1974), e.g. association of a flavour with a disease or a recovery from disease renders this flavour unpleasant or pleasant ; (v) and, finally, drugs (Fantino, Hosotte and Apfelbaum, 1986). Positive alliesthesia indicates a change to a more pleasurable sensation ; negative alliesthesia a change to a less pleasurable one.
Sensory pleasure and displeasure thus appear especially well suited as motivation for thermoregulatory behaviour (Attia, 1984). The case of pleasurable flavours shows an identical pattern (Fantino, 1984). Alimentary flavours are pleasurable during hunger and become unpleasant or indifferent during satiety. Measurement of human ingestive behaviour confirms the above relationship. Preference shows a qualitative influence : human subjects ingest more of what they like. It also shows a quantitative influence : the amount eaten is a function of the alimentary restrictions and increases after dieting, when negative alliesthesia has disappeared. The result is that pleasure scales can be used to judge the acceptability of food.

Conflicts : When placed in a conflict of biological motivations, e.g. fatigue in chest vs lower limbs, or fatigue vs thermal comfort, human subjects tend to maximize the sum of their sensory pleasures. In so doing they optimalize their

physiological functions and solve the problem of the conflicting motivations.
Fatigue is usually considered as a discriminative perception (Borg, 1962). This perception can also be studied experimentally from an affective (hedonistic) point of view. In experiments where subjects could select the intensity of their muscular exercise to accomplish a given task on a treadmill, they managed to work at constant power from one session to another (Figure 2), although the slope or the speed of the treadmill was varied by the experimenter. Such performance could be predicted from the rating by subjects of the affective sensations aroused in the chest and lower limbs. Optimalization was the result of minimization of sensory displeasure (Cabanac, 1985).

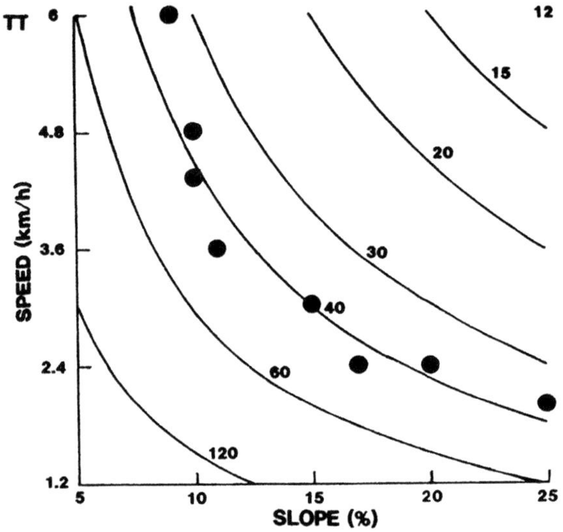

Fig. 2 Slope and speed chosen at the end of sessions by one subject when he was instructed to climb 300 m. In 4 sessions the treadmill slope was imposed by the experimenter and the subject could adjust the speed. In the 4 other sessions the treadmill speed was imposed by the experimenter and the subject could adjust the slope. Each dot stands for one session. The isochronic lines on the graph show the theoretical time, in minutes, necessary to climb 300 m (from Cabanac, 1985).

Fatigue was pitted against thermal discomfort in experiments where the subjects could adjust either ambient temperature or the intensity of muscular exercise. Here again the results showed optimalization of behaviour : behavioural thermogenesis was inversely proportional to ambient temperature (Figure 3). In the bidimensional space defined by all the ambient temperatures and all intensities of muscular exercise explored, the subjects' behaviour coincided with areas of sensory pleasure in the bidimensional sensory-space defined. Whichever muscular exercise, ambient temperature, or combinations of these two were presented, the subjects always optimized their behavior in maximizing sensory pleasure (Cabanac and LeBlanc, 1983).

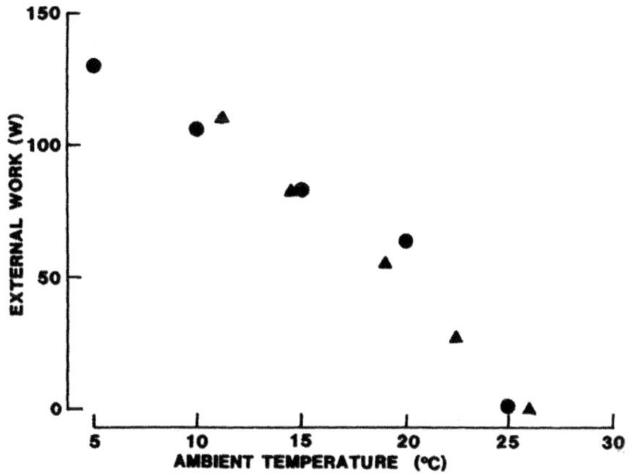

Fig. 3. Intensity of external muscular power produced by one subject at the end of one hour walking on a treadmill set at a constant speed (3 km.h^{-1}). External power was the work necessary for lifting the body mass, per unit of time; external power was only produced when treadmill slope 0. In a series of 10 sessions, either ambient temperature ( ) or treadmill slope ( ) was imposed by the experimenter and the other adjusted by the subject himself. Each point stands for one session (from Cabanac and LeBlanc, 1983). It can be seen that whether ambient temperature or treadmill slope was varied made no difference, all the points fall on the same line; the intensity of exercise chosen by the subject was proportional to ambient temperature, i.e. was a clearcut thermoregulatory behaviour.

Thus, the seeking of pleasure and avoidance of displeasure lead to homeostatic (short term), or homeorhetic (long term), behaviours. Pleasure, therefore, indicates a useful stimulus and simultaneously motivates the subject to enjoy it. Both a reward and a motivation, pleasure leads to optimization of life mechanisms.

Animals : Before examining how relevant all this is to animal behaviour and to farm animals, it should first be pointed out that in the animal kingdom the species less dependent on behaviour to achieve their physiological function are those that appeared latest. The development of the milieu intérieur and the perfection of physiological functions in mammals and birds has freed their behaviour from short-term physiological constraints and may explain their success in the competition of species. This is not to say that behaviour does not play a part in their physiology but just to recall that they draw one degree of liberty from their autonomic function while lower species, with the efficacious physiology, depend still more on behaviour. If the lower species rely exclusively on behaviour it can be expected that cognition must have a modest role in the determinism of physiological behaviour. On the other hand, the seeking of pleasure and the avoidance of displeasure may explain the adaptation of behaviour to its biological aims, without the implication of animal awareness. Often behaviour does not need cognition, e.g. postural adaptation to ambient temperature or food intake are displayed by anencephalic rats. Such a hypothesis would advantageously replace such an obscure concept as instinct. The animal would just seek to maximize pleasure and need not to know why.

The question then arises as to the possibilities of obtaining evidence that animals would seek sensory pleasure and be able to trade-off some amount of displeasure for this pleasure.

In the obstruction method, the strength of a motivation is measured not as a motor response or the consumption of a reward but rather the decision made by an animal to overcome a resistance and to obtain a reward (Warden, 1931). Such a situation can be explored in the laboratory under conditions close to natural conditions (Cabanac and Johnson, 1983). Rats were trained to feed daily from 10 a.m.to 12 noon. Once a week in an environment with an ambient temperature of - 15°C, -i.e. one extremely aggressive to rats- additional food was made available 16 m from a thermoneutral refuge. The additional food offered was varied in various sessions. Although laboratory chow was also always available in their thermoneutral home, the rats invariably ran out into the cold to the feeder, especially when the food was highly palatable. With such foods, the animals took as much as half their nutrient intake in the cold. For less palatable food, the rats went only once or twice to the feeder, and there stayed for only a short time. Thus the animals faced painful cold not from necessity, since food was provided in their warm shelter, but for the pleasure of ingesting a palatable bait. Such a result can be interpreted in the light of human experiments by stating that the rats exercized their behavioural degree of liberty to increase their sensory pleasure.

It is likely that the behaviours displayed by pigs, that preferred to feed rather than thermoregulate (Ingram & Legge, 1970), by steers that preferred to thermoregulate than feed, and by sheep that preferred to drink rather than feed (Squires & Wilson, 1971) are examples under farm conditions of animals maximizing their sensory pleasure and thus optimizing their behavioural response to stressful environments. It is therefore reasonable to let farm animals enjoy the environmental and social environment that they themselves seek (Dantzer, 1981 ; Wiepkema, 1985).

One basic postulate of ethology is that behaviour tends to satisfy the most urgent need of the behaving subject (Baerends, 1956 ; Tinbergen, 1950). Figure one shows, already, one possible trade-off between inflow and outflow. In addition, the simplified diagram incorporates in one pathway several possible behaviours. Wild animals are capable of coping with nutritional stress by substituting various responses (King & Murphy, 1985). Maximization of pleasure may be the link between physiology and behaviour and may explain how physiological optimization is achieved.

REFERENCES

Attia, M. 1984. Thermal pleasantness and temperature regulation in man. Neurosci. biobehav. Rev., 8, 335-342.

Baerends, G.P. 1956. Aufbau des tierischen Verhaltens. In : Handbuch der Zoologie (Ed. W. Kükenthal and T. Krumbach). (De Gruyter & Co, Berlin). Mammalia Teil. 10.

Borg, G. 1962. Physical performance and perceived exertion; Studia psychol. paedol., 11, 1-64.

Cabanac, M. 1971. Physiological role of pleasure. Science, 173, 1103-1107.

Cabanac, M. 1985. Optimisation du comportement par la minimisation du déplaisir dans un espace sensoriel à deux dimensions. C.R. Acad. Sc. Paris, 301(III), 607-610.

Cabanac, M. and Johnson, K.G. 1983. Analysis of a conflict between palatability and cold exposure in rats ; Physiol. Behav., 31, 249-253.

Cabanac, M. and LeBlanc, J. 1983. Physiological conflict in humans : fatigue vs cold discomfort. Am. J. Physiol., 244, R621-R628.

Cabanac, M. and Russek, M. 1982. Régulation et contrôle en biologie (Presses de l'Université Laval, Québec).

Dantzer, R. 1981. Le stress des animaux d'élevage. La recherche, 12, 280-289.

Fantino, M. 1984. Role of sensory input in the control of food intake. J.auton. nerv. Syst., 10, 347-358.

Fantino, M., Hosotte, J. and Apfelbaum, M. 1986. An opioide antagonist, naltrexone, reduces the preference for sucrose in man. Amer. J. Physiol. (in press).

Garcia, J. Hankins, W.G. and Rusiniak, K.W. 1974. Behavioral regulation of the milieu interne in man and rat. Science, 185, 824-831.

Ingram, D.L. and Legge, K.F. 1970. The thermoregulatory behavior of young pigs in a natural environment; Physiol. Behav., 5, 981-987.

King, J.R. and Murphy, M.E. 1985. Periods of nutritional stress in the annual cycles of endotherms; Fact or fiction ? Amer. Natur., 25, 955-964.

Malechek, J.C. and Smith, B.S. 1976. Behavior of range cows in response to winter weather. J. Range Management, 29, 9-12.

Marks, L.E. and Gonzalez R.R. 1974. Skin temperature modifies the pleasantness of thermal stimuli. Nature, 247, 473-475.

Nicolaïdis, S. 1977. Physiologie du comportement alimentaire. In:Physiologie Humaine (ed. Ph. Meyer) (Flammarion, Paris). pp.908-922.

Pfaffmann, C. 1960. The pleasures of sensation. Psychol. Rev., 67, 253-268.

Rolls, B.J., Rowe, E.A., Rolls, E.T., Kingston, B., Megson, A. and Gunary, R. 1981. Variety in a meal enhances food intake in man. Physiol. Behav., 26, 215-221.

Rolls, B.J., Wood, R.J., Rolls, E.T., Lind, H., Lind, W. and Ledingham, J.G.G. 1980. Thirst following water deprivation in humans. Amer. J. Physiol., 239, R476-R482.

Squires, V.R. and Wilson, A.D. 1971. Distance beween food and water supply and its effect on drinking frequency and food and water intake of merino and border Leicester sheep. Austr. agricult. Res., 22, 283-290.

Tinbergen, N. 1950. The hierarchical organization of mechanisms underlying instinctive behaviour. Sympos. Soc. exper. Biol., 4, 305-312.

Toates, F. 1986. Motivational systems (Cambridge University Press, Cambridge.)

Warden, C.J. 1931. Animal motivation, Experimental studies on the Albino Rat. (Columbia University Press, New York).

Wiepkema, P.R. 1985. Abnormal behaviours in farm animals : ethological implications. Nether. J. Zool., 35, 279-299.

Wundt, W. 1874. Grundzüge der physiologischen Psychologie (Engleman, Leipsig).

Young, P.T. 1959. The role of affective process in learning and motivation. Psychol. Rev., 66, 104-125.

LIST OF PARTICIPANTS

BELGIUM

Dr. Y. LAMBERT
Biologie du Comportement
Université Catholique de Louvain
1, Croix du Sud
B - 1348 Louvain-La-Neuve

Dr. F.O. ÖDBERG
Leerstoel voor Dierlijke Genetica en Veeteelt
Rijksuniversiteit Gent
Heidestraat 19
B - 9220 Merelbeke

DENMARK

Dr. K.E. HELLER
Institute of Population Biology
Copenhagen University
Universitetsparken 15
DK - 2100 Copenhagen

Dr. G. NØRGAARD-NIELSEN
Royal Veterinary and Agricultural University
Department of Forensic and State Veterinary Medicine
13, Bülowsvej
DK - 1870 Frederiksberg C

Dr. K. VESTERGAARD
Royal Veterinary and Agricultural University
Department of Forensic and State Veterinary Medicine
13, Bülowsvej
DK - 1870 Frederiksberg C

FEDERAL REPUBLIC OF GERMANY

Dipl. Ing. Agr. E. VON BORELL
Institut für Tierzucht und Tierverhalten-FAL
Institutsteil Trenthorst/Wulmenau
D - 2061 Westerau 2

Dr. K. DÄMMRICH
Institut für Veterinär-Pathologie
Freie Universität Berlin
Strasse 518, No. 15
D - 1000 Berlin 37

Dr. Dr. habil. H. EICHINGER
Versuchsstation Thalhausen
Lehrstuhl für Tierzucht
Technische Universität München
Weihenstephan
D - 8051 Kranzberg

Dr. J. LADEWIG

Institut für Tierzucht und
Tierverhalten-FAL
Institutsteil Trenthorst/Wulmenau
D - 2061 Westerau 2

FRANCE

Dr. M. CABANAC

Universite Claude-Bernard
Faculte de Medecine Lyon Sud
Laboratoire de Physiologie
B.P. 12
69921 Oullins Cedex

Dr. R. DANTZER

Institut National de la Sante et
de la Recherche Medicale
Unite 259
Domaine de Carreire
Rue Camille-Saint-Saëns
33077 Bordeaux

Dr. J.M. FAURE

Ministere de l'Agriculture
Institut National de la Recherche
Agronomique
Centre de Tours
Nouzilly 37380 Monnaie

Dr. P. MORMEDE

Institut National de la Sante et
de la Recherche Medicale
Unite 259
Domaine de Carreire
Rue Camille-Saint-Saëns
33077 Bordeaux

ITALY

DR. A. OLIVERIO

C.N.R. Istituto di Psicobiologia e
Psicofarmacologia
Via Reno, 1
00198 Roma

Dr. M. VERGA

Istituto di Zootecnica Veterinaria
Universita di Milano
Via Celoria, 10
20133 Milano

THE NETHERLANDS

DR. P.W.M. VAN ADRICHEM

Agricultural University
Department of Animal Physiology
Haarweg 10
6709 PJ Wageningen

Dr. R.E. BALLIEUX
Academisch Ziekenhuis Utrecht
Universiteitskliniek Inwendige Geneeskunde
Afdeling Immunologie
Catharijnesingel 101
3511 GV Utrecht

Dr. B. BOHUS
Department of Animal Physiology
Biological Centre
P.O. Box 14
9750 AA Haren

Dr. G. BEUVING
C.O.V.P. Spelderholt
Spelderholt 9
7361 DA Beekbergen

Ir. H.J. BLOKHUIS
C.O.V.P. Spelderholt
Spelderholt 9
7361 DA Beekbergen

Ir. R. DONKER
Agricultural University
Department of Animal Husbandry
Marijkeweg 40
6709 PG Wageningen

Dr. J.M. KOOLHAAS
Deparment of Animal Physiology
Biological Centre
P.O. Box 14
9750 AA Haren

Dr. B. OLIVIER
Department of Pharmacology
Duphar
P.O. Box 2
1380 AA Weesp

DR. J.M. VAN REE
Rudolf Magnus Institute
Department of Pharmacology
Medical Faculty Utrecht
Vondellaan 6
3521 GD Utrecht

Dr. W.G.P. SCHOUTEN
Agricultural Univerisity
Department of Animal Husbandry
Marijkeweg 40
6709 PG Wageningen

Dr. W. SYBESMA
IVO Zeist
P.O. Box 501
3700 AM Zeist

Dr. P.R. WIEPKEMA
Agricultural University
Department of Animal Husbandry
Marijkeweg 40
6709 PG Wageningen

REPUBLIC OF IRELAND

Dr. P.V. TARRANT	The Agricultural Institute Dunsinea Research Centre Castleknock Dublin 15
Dr. J.F. ROCHE	Faculty of Veterinary Medicine University College Dublin Ballsbridge Dublin 4

UNITED KINGDOM

Dr. B.A. BALDWIN	AFRC, Institute of Animal Physiology Babraham Cambridge
Dr. D.M. BROOM	University of Cambridge Dept. Clinical Veterinary Medicine Madingley Road Cambridge CB3 OE5
Dr. V.G. KITE	AFRC Poultry Research Centre Roslin Midlothian EH25 9PS Scotland
Dr. F.M. TOATES	Bioloy Department The Open University Walton Hall Milton Keynes MK7 6AA
Dr. D.M. WARBURTON	University of Reading Department of Psychology Whiteknights Reading RG6 2AL

UNITED STATES

Dr. H.S. SIEGEL	Department of Poultry Science Pennsylvania State University 213 W.L. Henning Building University Park Pennsylvania 16802

CEC

Mr. J. CONNEL	Commission of the European Communities 200, Rue de la Loi B-1049 Bruxelles

GPSR Compliance

The European Union's (EU) General Product Safety Regulation (GPSR) is a set of rules that requires consumer products to be safe and our obligations to ensure this.

If you have any concerns about our products, you can contact us on ProductSafety@springernature.com

In case Publisher is established outside the EU, the EU authorized representative is:

Springer Nature Customer Service Center GmbH
Europaplatz 3
69115 Heidelberg, Germany

Batch number: 10755292

Printed by Printforce, the Netherlands